Climate in Context

Climate in Context: Science and Society Partnering for Adaptation

EDITED BY

Adam S. Parris
Science and Resilience Institute at Jamaica Bay, Brooklyn College,
2900 Bedford Ave, Brooklyn, NY 11210, USA

Gregg M. Garfin
Climate Assessment for the Southwest (CLIMAS), School of Natural Resources and the Environment,
Institute of the Environment, The University of Arizona, 1064 E. Lowell St., Tucson, AZ 85721, USA

Kirstin Dow
Carolinas Integrated Sciences and Assessments RISA, Department of Geography,
University of South Carolina, 709 Bull Street, Columbia, SC 29208, USA

Ryan Meyer
California Ocean Science Trust, 1330 Broadway, Suite 1530, Oakland, CA 94612, USA

Sarah L. Close
University Corporation for Atmospheric Research, in service to: Climate and Societal Interactions Division,
NOAA Climate Program Office, 1315 East-West Highway, SSMC3, Silver Spring, MD 20910, USA

American Geophysical Union

Library of Congress Cataloging-in-Publication Data applied for.

ISBN: 9781118474792

A catalogue record for this book is available from the British Library.

Wiley also publishes its books in a variety of electronic formats. Some content that appears in print may not be available in electronic books.

Cover image: Getty/Åsa Almer/EyeEm

Typeset in 10/13.5pt MeridienLTStd by SPi Global, Chennai, India
Printed and bound in Singapore by Markono Print Media Pte Ltd

1   2016

Contents

List of contributors

Daniel Bader

Center for Climate Systems Research, Columbia University Earth Institute, 2880 Broadway, New York, NY 10025, USA

NASA Goddard Institute for Space Studies, 2880 Broadway, New York, NY 10025, USA

Michael Beaulac

Michigan Department of Environmental Quality, Executive Division, 525 West Allegan St., Lansing, MI 48909-7973, USA

Stuart Blythe

Writing, Rhetoric and American Cultures, Michigan State University, 220 Trowbridge Rd, East Lansing, MI 48824, USA

David Brown

National Centers for Environmental Information, National Oceanic and Atmospheric Administration, Fort Worth, TX 76102, USA

Timothy J. Brown

California and Nevada Applications Program (CNAP), Desert Research Institute, 2215 Raggio Parkway, Reno, NV 89511, USA

Julie Brugger

Institute of the Environment, University of Arizona, Tucson, AZ 85721-0137, USA

Lorraine Cameron

Michigan Department of Community Health, Division of Environmental Health, 201 Townsend St., Lansing, MI 48913, USA

Greg Carbone

Carolinas Integrated Sciences and Assessments RISA, Department of Geography, University of South Carolina, 709 Bull Street, Columbia, SC 29208, USA

Sarah L. Close

University Corporation for Atmospheric Research, in service to: Climate and Societal Interactions Division, NOAA Climate Program Office, 1315 East-West Highway, SSMC3, Silver Spring, MD 20910, USA

Michael Crimmins
Climate Assessment for the Southwest, Department of Soil Water and Environmental Science, University of Arizona, Tucson, AZ 85721-0038, USA

Art DeGaetano
Earth and Atmospheric Science and Northeast Regional Climate Center, Cornell University, Ithaca, NY 14850, USA

Ed Delgado
Bureau of Land Management (BLM), National Interagency Coordination Center, National Interagency Fire Center, 3833 S. Development Ave., Boise, ID 83705, USA

Lisa Dilling
Western Water Assessment, Cooperative Institute for Research in Environmental Sciences, University of Colorado Boulder, 216 UCB, Boulder, CO 80309, USA
Environmental Studies Program and Center for Science and Technology Policy Research, Cooperative Institute for Research in Environmental Sciences, University of Colorado Boulder, 4001 Discovery Drive, Boulder, CO 80309-0397, USA

Kirstin Dow
Carolinas Integrated Sciences and Assessments RISA, Department of Geography, University of South Carolina, 709 Bull Street, Columbia, SC 29208, USA

Daniel B. Ferguson
Institute of the Environment, University of Arizona, 1064 E. Lowell Street, Tucson, AZ 85721, USA

Melissa L. Finucane
East West Center, 1601 East-West Rd, Honolulu, HI 96848, USA
RAND Corporation, 4570 Fifth Ave #600, Pittsburgh, PA 15213, USA

Jay Fowler
Department of Geography, Carolinas Integrated Sciences and Assessments, University of South Carolina, 709 Bull Street, Columbia, SC 29208, USA

Clyde Fraisse
Southeast Climate Consortium, Department of Agricultural and Biological Engineering, University of Florida, Gainesville, FL 32611-0570, USA

J. Brook Gamble
Alaska Center for Climate Assessment and Policy, University of Alaska Fairbanks, 505 S Chandlar Drive, Fairbanks, AK 99775, USA

Gregg Garfin
Climate Assessment for the Southwest (CLIMAS), School of Natural Resources and the Environment, Institute of the Environment, University of Arizona, 1064 E. Lowell St., Tucson, AZ 85721, USA

Eric S. Gordon
Western Water Assessment, Cooperative Institute for Research in Environmental Sciences, University of Colorado Boulder, 216 UCB, Boulder, CO 80309, USA

Holly Hartmann
University of Arizona, Tucson, AZ 85721, USA

Radley Horton
Center for Climate Systems Research, Columbia University Earth Institute, 2880 Broadway, New York, 10025, NY, USA
NASA Goddard Institute for Space Studies, 2880 Broadway, New York, NY 10025, USA

Nathan P. Kettle
Alaska Center for Climate Assessment and Policy, University of Alaska Fairbanks, and Alaska Climate Science Center, 505 S Chandlar Drive, Fairbanks, AK 99775, USA

Victoria W. Keener
East West Center, 1601 East-West Rd, Honolulu, HI 96848, USA

Kirsten J. Lackstrom
Carolinas Integrated Sciences and Assessments RISA, Department of Geography, University of South Carolina, 709 Bull Street, Columbia, SC 29208, USA

Ellen Lay
University of Arizona, Tucson, AZ 85721, USA

Maria Carmen Lemos
Great Lakes Integrated Sciences and Assessments RISA, School of Natural Resources and Environment, University of Michigan, 430 E. University Ave, Ann Arbor, MI 48109-1115, USA

Ralph Levine
Department of Community Sustainability, Michigan State University, 220 Trowbridge Rd, East Lansing, MI 48824, USA

Elizabeth McNie
Western Water Assessment, Cooperative Institute for Research in Environmental Sciences, University of Colorado Boulder, 216 UCB, Boulder, CO 80309, USA

Chad McNutt
National Integrated Drought Information System Program Office, University Corporation for Atmospheric Research, Boulder, CO 80307-3000, USA

Ryan Meyer
California Ocean Science Trust, 1330 Broadway, Suite 1530, Oakland, CA 94612, USA

Laura Schmitt Olabisi
Department of Community Sustainability, Michigan State University, 220 Trowbridge Rd, East Lansing, MI 48824, USA

Gigi Owen
Institute of the Environment, University of Arizona, 1064 E. Lowell Street, Tucson, AZ 85721, USA

Adam Parris
Science and Resilience Institute at Jamaica Bay, Brooklyn College, 2900 Bedford Ave, Brooklyn, NY 11210, USA

Andrea J. Ray
Western Water Assessment, Cooperative Institute for Research in Environmental Sciences, University of Colorado Boulder, 216 UCB, Boulder, CO 80309, USA
Physical Sciences Division, NOAA Earth System Research Laboratory, 325 Broadway, R/PSD1, Boulder, CO 80305, USA

Jinyoung Rhee
APEC Climate Center, 12, Centum 7-ro, Haeundae-gu Busan 48058, South Korea

Rachel E. Riley
Southern Climate Impacts Planning Program RISA, Oklahoma Climatological Survey, University of Oklahoma, 120 David L. Boren Blvd., Suite 2900, Norman, OK 73072, USA

Cynthia Rosenzweig
Center for Climate Systems Research, Columbia University Earth Institute, 2880 Broadway, New York, NY 10025, USA
NASA Goddard Institute for Space Studies, 2880 Broadway, New York, NY 10025, USA

Mark Shafer
Southern Climate Impacts Planning Program, University of Oklahoma, 120 David L. Boren Blvd., Suite 2900, Norman, OK 73072, USA

Caitlin F. Simpson
U.S. Department of Commerce, NOAA Climate Program Office,
1315 East West Highway, Room 12212, Silver Spring, MD 20910, USA

Linda Sohl

Center for Climate Systems Research, Columbia University Earth Institute, 2880 Broadway, New York, 10025, NY, USA

NASA Goddard Institute for Space Studies, 2880 Broadway, New York, NY 10025, USA

William Solecki

Department of Geography, CUNY Institute for Sustainable Cities, Hunter College, 695 Park Avenue, New York, NY 10065, USA

John Stevenson

Climate Impacts Research Consortium, Oregon Sea Grant, Oregon State University, Corvallis, OR 97331-5503, USA

Sarah F. Trainor

Alaska Center for Climate Assessment and Policy, University of Alaska Fairbanks, 505 S Chandlar Drive, Fairbanks, AK 99775, USA

Robert S. Webb

Physical Sciences Division, NOAA Earth System Research Laboratory, 325 Broadway, R/PSD1, Boulder, CO 80305, USA

Jessica Whitehead

Carolinas Integrated Sciences and Assessments, North Carolina State University, North Carolina Sea Grant, Raleigh, NC 27695-8605, USA

Tom Wordell

(Retired), USDA Forest Service, National Interagency Coordination Center, National Interagency Fire Center, 3833 S. Development Ave., Boise, ID 83705, USA

Foreword

From the beginning, the Regional Integrated Sciences and Assessments (RISA) program has been an experiment. Unlike other experiments in climate-related services and in connecting science with decision-making, it has survived and thrived since 1995, despite numerous challenges. A partnership between the National Oceanic and Atmospheric Administration (NOAA), universities and stakeholders, RISA is focused on place-specific problems and solutions and explores the space between research and decision-making. It is designed to respond in a flexible way to the chain of requests for climate-related information at multiple time scales that has risen from regions and sectors across the United States.

RISA's origins lie in the vision of a few important leaders in the federal government and in academia; they include J. Michael (Mike) Hall, the former director of the Office of Global Programs at NOAA and a widely influential visionary in the U.S. government; Edward (Ed) Miles, the director of the first RISA, the Climate Impacts Group at the University of Washington; and Claudia Nierenberg, who served as the first program director for RISA. Of course, there have been literally dozens of visionary leaders associated with the program since the early days, but without the contributions of these three people, the program would not be what it is today. A strong vision from the beginning, a flexible management approach, and strong central coordination and leadership have all played a major role in building the program.

The 11 RISA program "experiments" across the country are linked to a multi-institutional network managed centrally in the Climate Program Office at NOAA. Each RISA has evolved in response to the interests and capabilities of Principal Investigators (PIs) within the partner universities as well as to the interests and needs of its regional stakeholders. Because there are so many issues related to climate impacts, vulnerability assessment, and building relevant decision-support products over a range of time and space scales, there is a need for multiple different approaches across the United States.

Clearly, the seed funding provided by NOAA has provided incentives to take the experiments in particular directions, but in virtually all cases, a highly leveraged program has evolved that includes a range of different local, regional, and federal funding sources and partnerships. This flexibility and diversity is one of the institutional strengths of RISA. Although the program has always been underfunded, RISA is widely acknowledged as a success story in providing decision-relevant science products. It has been a constant challenge to maintain and/or grow the program over time. Program funding has been threatened for a variety of internal (agency) and external reasons, but it has continued to provide incentives for interdisciplinary and transdisciplinary work that has been truly groundbreaking. In some cases, the interdisciplinary teams have been together for almost 20 years, and in all cases the opportunity to do longitudinal studies and engage with stakeholders over years-to-decades has been instrumental in building an understanding of both the art and science of connecting science and decision-making.

The fact that each of the RISAs has different topical focus areas and different strengths in engaging with stakeholders is an important part of the success of RISA as a system, and there are many examples of how the system itself has evolved as an institution over time. Not only are there far more collaborative projects across the RISA network now than there were historically, but there are also purposeful efforts to design the research in ways that fill both social and physical science gaps in the whole network. This approach is unprecedented in federal science programs and likely has no parallel globally, though there are now many examples of science networks that emulate portions of this approach.

Although there is an ongoing debate within the RISA community about where on the "science-to-action" continuum the work should focus, part of the rationale has always been to experiment with the space between research and applications, building an understanding of the role of science in policy and the role of academia and government investments in building science-based solutions. In building local communities of practice that are linked to a truly functional network of practitioners, there have been contributions to the careers of researchers and students who have been funded by the program as well as to the careers of external stakeholders who have been drawn into the experiment.

Building communication and planning tools, reporting back to the larger RISA community about successes and failures, and a significant dose of self-reflection have been the hallmarks of the program. In fact, self-reflection has been strongly encouraged. This willingness to openly expose weaknesses and identify the needs for improvement is extremely unusual in government-sponsored programs. There has been significant stress underlying all of the progress that has been made, but in many cases that stress has provided for enhanced learning opportunities (see also, RISA in a Nutshell).

RISA contributions in the national and global context: climate services, assessments, and adaptation

Given the well-documented observation that climate variability and climate change are already causing costly damages in every region and sector of the United States and the globe, and that there are many unrealized economic opportunities as well (National Climate Assessment, 2014), it has been clear to most of the climate community for two decades that climate information services at multiple time scales (subseasonal to interannual to decadal and beyond) are needed. The need for climate services mirrors the need for weather information but with longer time scales and larger consequences. Despite this fact, building a U.S. climate service has been very controversial. Without explicitly intending to do so, RISA has emerged as one of several highly leveraged attempts to fill in the gaps associated with the lack of a climate service in the United States. It also provides a good model for regional climate service activities that could be developed by other countries.

RISA has already influenced thinking outside of the U.S. borders through contributions of many of its PIs and stakeholders to the International Panel on Climate Change reports, through international colleagues who have closely followed the development of RISA-funded knowledge production, and through investments by the United States in international adaptation/resilience initiatives. For example, the influence of RISA contributions can be seen in the reports of the United States to the UN Framework Convention on Climate Change, the development of the Global Framework for Climate Services and the September 23, 2014 Presidential Executive Order on Climate Resilient International Development. Lessons learned have also influenced the evolution of the International Research Institute for Climate and Society, the Inter-American Institute for Global Change Research, and most recently, the framing of the NOAA International Research and Applications Program.

Many of the social science and process findings of RISA mirror those of other climate-focused organizations elsewhere, including the Commonwealth Scientific and Industrial Research Organization (CSIRO), Australian Government Science Program and the UKCIP (the UK Climate Impacts Program). Examples are: the need for building trusted relationships between scientists and stakeholders over time, building salience, credibility and legitimacy of products and processes in partnership with stakeholders, enhancing utility of information through co-production of knowledge, and boundary spanning activities. In addition, RISA-related findings have been extremely visible in advice to the US Global Change Research Program (USGCRP) and in multiple NRC panels (there are at least 15 reports where

RISA players and findings were influential, including the five America's Climate Choices reports).

The RISA network has also made very explicit contributions to all three US National Climate Assessments, particularly certain Synthesis and Assessment products (e.g., SAP 5.3) of the Bush Administration, and the regional components of the Third National Climate Assessment (NCA3). The contributions to the NCA3 were very substantial; RISA program managers, PIs, staff, and students played a role in virtually all of the eight regional assessment teams as well as in developing technical support documents and regional foundation reports later published as a book series by Island Press. All of the RISAs have been engaged in the discussions of the sustained approach to assessment and several are members of the NCA engagement network.

Science support for adaptation/resilience decisions is in great demand. Most of what the RISAs do, whether it is in support of agriculture, fire management, urban planning or water resources management, contributes to resilience— mostly direct investments in applied science, adaptation tools and services, climate-related communications, science translation in support of specific sectoral interests, social network analysis, and analysis of the effectiveness of decision support.

The interagency context: subsequent science networks and the needs and aspirations of federal agencies

As the costs of addressing climate-related disasters have risen, and the understanding of the climate-related drivers of these events have become more clearly understood, there has been an overall shift toward enhancing the societal relevance of the $2.6 billion annual research agenda of the 13-agency USGCRP. Although the proportion of funding that goes into decision support is very small (certainly less than 5%), the visibility of these efforts has continued to rise. Most of the movement in this direction has occurred within the last 6 years, though National Academy reports have been pushing the federal science agencies in this direction for decades and the Global Change Research Act of 1990 clearly anticipated that this transition would occur more rapidly. The lessons learned by the federal government in this arena have come in large part from lessons learned from RISAs.

Over the past decade, the demand for climate-related support from the federal government has far outstripped the capacity of the RISA system to meet the demand for decision support. Eight Climate Science Centers and 22 Landscape Conservation Cooperatives (supported by the Department of the Interior) and 6 new "regional climate hubs" (the U.S. Department of Agriculture) mean that the capacity for engaging with stakeholders and providing relevant science support is expanding. Each of these networks has a different focus and audience, but all are dependent on using the foundational science and engagement strategies employed by RISA. Clearly, there is enough potential for better outcomes and collaboration across these systems; all three networks will benefit if they work more closely together in the future.

Conclusions

Having provided some perspective on the history and achievements of the RISA program, we believe that RISA was both *a strategic* and an opportunistic response, an approach whose time was right in the mid-1990s but whose contributions are even more relevant today. RISAs are innovative experiments in decision-relevant science and in self-conscious evaluation of engagement processes. Their contributions extend far beyond the RISAs themselves to a much broader knowledge network, including colleagues and practitioners that now include former students, stakeholders, and collaborators across the United States and globally.

RISA has been a public good, providing value to U.S. citizens for 20 years … connecting the best available science produced through government agencies and academia to real world issues like drought planning, range management, agricultural production, water management, flood control, and recovery from disasters in coastal and urban settings. These contributions span the range from fundamental physical and social science to stakeholder engagement to decision support at multiple time and space scales. The intellectual and public policy outcomes of RISA now have global implications and are fundamental to major shifts in U.S. domestic and international policies.

We would like to acknowledge the efforts of all who have contributed to RISA in the past, and those who will continue to contribute, in helping to build a more resilient future.

Kathy Jacobs
Jim Buizer
The University of Arizona

Preface

"The local- to global-scale impacts of climate variability and change, as well as the broader issue of global change, have fueled a growing public demand for timely and accessible information about present and future changes … that can be applied directly to planning, management, and policymaking."

The National Global Change Research Plan 2012–2021 [1]

Usable science is an enduring concept, but the practice remains elusive. It happens in muddy waters, where the controlled space of basic science meets the ever-evolving context of people and places. After 20 years of exploration through the RISA program, this book is an attempt to make the practice of usable science more transferrable by explaining key techniques.

Climate in Context documents the mechanics of fostering engaged and collaborative approaches between researchers and practitioners to inform decision-making. The book is organized around four themes of the RISA program: Understanding Context and Risk, Managing Knowledge-to-Action Networks, Innovating Services, and Advancing Science Policy. These themes represent the steps in a process of developing usable science to address persistent and/or emerging climate-environment-society management issues. Experience tells us that some of these steps may occur simultaneously, or even out of the aforementioned sequence. These themes relate to well-known advice from sources such as the National Research Council's report, "Informing Decisions in a Changing Climate" [2]. However, Climate in Context demonstrates that RISA teams have moved from articulating a course of implicit learning—a passive benefit of the interactions between researchers, intermediates, and decision-makers—to one of explicit learning, where the social capitals for learning and communication across social networks [3] are actively engaged at the outset of new projects. Leveraging the learning process is what allows researcher-practitioner partners to move from informing decisions with data and information to a keystone societal capacity—the knowledge and policy to manage risk in an increasingly uncertain world. Almost every chapter in *Climate in Context* refers to some form of learning as an essential element in understanding decision contexts, setting expectations for product development, and coproducing the knowledge needed to increase preparedness for the impacts of climate variability and change.

Unlike data and information, knowledge is not a commodity. The case studies and syntheses in *Climate in Context* show that usable science can generate the type of knowledge desired for adaptive planning and action. As shown in the chapters in this book, iterative and adaptive learning must accompany the transfer of technology and data, and it requires diverse approaches. Within the bounds of federal agency protocol, the RISA program has allowed "a thousand flowers to bloom" among the regional teams, in order to learn from the multiple geographic, political, and institutional contexts across the United States. The authors contributing to *Climate in Context* demonstrate, time and again, that partnerships are a central part of the practice of usable science. Climate science researchers aim to understand the dynamics and statistics of the climate system and develop products to support decisions, whereas practitioners seek knowledge to address time-sensitive decisions in a cost-effective manner. Fostering partnerships and managing the exchange of knowledge across the boundaries between researchers and practitioners require guidance and knowledge from social scientists and facilitators whose investigations on decision context lead the way to fostering and replicating innovation and learning. But how, specifically, do you achieve this process? When do you employ which technique? In what combination? And how do you know when you have achieved this? The answers to these questions are the substance of *Climate in Context*.

Members of the first RISA team—the Climate Impacts Group—used the phrase "a voyage of discovery" to describe this process of adaptive learning. The scientists whose contributions make up this book are some of the many voices of a growing field that is ushering further discovery. As detailed in the foreword, there is unmet demand for greater capacity in usable science. The demand reflects a modest notion that, despite its far-reaching impact, there remains a great deal of unmet potential in the ability of science to provide public value. In the face of increasing risk, society could easily retrace well-worn paths to reactive responses to climate-related disasters—costing needless loss of lives and extra billions of dollars; however, two decades of learning in the RISA program has taught us to apply usable science to proactively build capacity and increase preparedness to climate-related risks. We caution the reader: as is often the case with good intentions, the pursuit of usable science has been both controversial and prone to folly. *Climate in Context* is not an attempt to accelerate the application of rote processes, but rather to accelerate adaptive and collaborative learning to promote the expansion of usable science. It is an invitation to use these techniques and lessons learned when needs, context, and scientific insights align.

On behalf of the editors, Adam Parris and Gregg Garfin

References

1 U.S. Global Change Research Program (2012) *The National Global Change Research Plan 2012–2021*. A strategic plan for the U.S. Global Change Research Program. National Coordination Office for the U.S. Global Change Research Program, Washington, DC, pp. 132.

2 National Research Council (NRC) (2009) *Informing Decisions in a Changing Climate*. Panel on Strategies and Methods for Climate- Related Decision Support. Committee on the Human Dimensions of Global Change, Division of Behavioral and Social Sciences and Education. National Academies Press, Washington, DC, pp. 188.

3 Brown, P.R., Nelson, R., Jacobs, B. *et al.* (2010) Enabling natural resource managers to self-assess their adaptive capacity. *Agricultural Systems*, **103**, 562–568.

Acknowledgments

Like its subject, this book is a community effort. The development of the book alone has spanned a few natural disasters, a presidential election, the third National Climate Assessment, and a change in NOAA leadership. Editors and authors have changed jobs, started families, finished graduate school, and moved to new regions. Among the editors, we include the able and patient editorial and production staff of Wiley & Sons, Inc., especially Rachael Ballard, with whom we first developed the idea for a book on the RISA program. Taken in sum, this book is a testament to the collective endurance of a community of practice, within and outside the RISA program. Funding and management from NOAA remain a critical element of this endurance.

Ryan Meyer and Kirstin Dow deserve special recognition for their wise and patient contributions to the editorial working group, which has overseen the life of the project. Their devotion to exceptional editorial rigor improved all of the chapters. Sarah Close provided critical input and dedication in the tough final stages. All of the authors showed dedication and persistence, and we gratefully acknowledge the many experts who provided external peer review. The book does not capture many worthy and important RISA efforts, some of which have been formally submitted as chapters; we thank those authors for devoting time to the book at its nascent stage as well as to the RISA program. We also wish to recognize all of the scientists and experts who have contributed to RISA, both directly and indirectly, including the pioneers in the Climate Impacts Group at the University of Washington.

The foreword to this book rightfully acknowledges several key individuals for their foundational efforts in establishing RISA, though, by dint of modesty, fails to mention the substantial contributions of James Buizer. We wish to dedicate this book to two of those individuals, J. Michael Hall and Edward L. Miles.

Disclaimer

The scientific results and conclusions, as well as any views or opinions expressed herein, are those of the authors and do not necessarily reflect the views of NOAA or the Department of Commerce

Background on RISA

Twenty years ago, a team of researchers began what would become a series of projects aimed at understanding how climate information can better inform decisions to adapt to a changing environment. It began with the establishment of the first Regional Integrated Sciences and Assessments (RISA), Climate Impacts Group (CIG) at the University of Washington in 1995. From there, the National Oceanic and Atmospheric Administration's (NOAA's) RISA program, managed from NOAA's Climate Program Office (formerly the Office of Global Programs), has grown to a network of 11 regionally based teams (Figure. I.1) around the United States. To this day, the RISA program supports interdisciplinary research teams that help expand and build the capacity of those seeking to prepare for and adapt to climate variability and change.

In the early stages of the program, RISA teams were developed through deliberate consultations between RISA program management at NOAA and university-based scientists around specific issues or focusing events in a region. In 2006, the program transitioned to a competitive funding model in which each region is competed on a 5-year cycle. RISA teams developed diverse research areas and management structures, allowing for experimentation and learning across the network. Throughout the 20-year history of the network, the hallmarks of RISA work have been a central commitment to partnerships, process, and building trust.

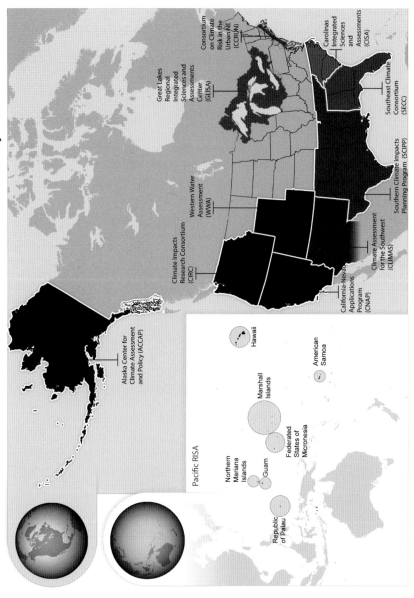

Figure I.1 The 11 RISA teams and their geographic footprints, as of February 2015.

SECTION I

Understanding context and risk

A central, enduring finding over the 20 years of the Regional Integrated Sciences and Assessments (RISA) program is that climate information can inform decisions to adapt to a changing environment, but only if the climate research community and decision-makers work together to understand each other's priorities, needs, and capabilities. This volume brings together insights and lessons on approaches to support climate decisions drawn from the experience of RISA teams and from collaborative dialogues among teams and partners. It begins by exploring, in this first theme entitled Understanding Context and Risk, the complex and evolving science needs embedded within decision contexts not exclusively related to climate, but responsive to other drivers of environmental, institutional, and societal change.

RISA experiences suggest that the task of identifying and refining climate science needs can often take years. In addressing this multidimensional issue, RISA investigators work to distinguish between what constitutes an initially perceived need and a more specific, grounded need for better comprehension of options within a range of uncertain futures. The first questions posed by decision-makers often become more precise during discussions of near-term scientific possibilities and limitations.

RISA teams build and sustain a capability for understanding emerging and evolving decision contexts informed by both advancing knowledge of how people use scientific information in the decision-making process and emerging information about projected exposure to climate-related impacts. The first chapter in this theme, by Simpson *et al.*, provides a broad overview of the capabilities developed by RISA teams to inform decisions about when to invest or redistribute research resources to meet needs within these changing contexts. They assess the multiple social science methods employed by RISA teams to build the foundation for coproduction of knowledge and policy. Their chapter highlights the evolution of research approaches needed as ongoing working relationships between RISA teams and practitioners deepen over time; these include reviewing documents, conducting surveys and interviews, participant observation and, more recently, investigating informal institutions, such as social and knowledge networks. For science to be actionable, it must be salient, timely, and credible. Ray and Webb reflect on the use of a time-honored framework, the decision calendar, to establish mutual understanding of the timely insertion of assessment and forecast information in climate-related decisions.

While RISAs are credited for their ability to engage decision-makers and to understand context, the complementary capacities to develop risk-based information and to support knowledge exchange often receives less recognition. Horton *et al.* show how an event such as Hurricane Sandy warranted re-allocation of time and expertise to develop risk information in support of greater resilience in an urban rebuilding strategy. In addition, the RISA follow-up work on Hurricane Sandy is intended to complement ongoing research on the exposure and vulnerability of a regional urban system to regional climatic and nonclimatic stressors, such as increasing water demands. They point out the challenges of keeping multiple risks in perspective, given urban planners' visceral experience of the hurricane, and illustrate how a RISA team advanced a better understanding of vulnerability and risk in the form of usable knowledge for New York City, ultimately building a capacity to adapt to global change.

Understanding contexts and risks, which take many institutional and methodological forms, requires expertise in forming and managing interactions among scientists and decision-makers, and among scientists from different disciplines, a topic that is further explored in Managing Knowledge-to-Action Networks. Working relationships and collaborative partnerships are established through the exchange of knowledge. In many instances, RISA social scientists consider the dynamics of these interactions and raise important questions—highlighted in the chapters of Understanding Context and Risk—relevant to expanding the scientific community's capacity to more directly serve society.

CHAPTER 1

Assessing needs and decision contexts: RISA approaches to engagement research

Caitlin F. Simpson[1], Lisa Dilling[2], Kirstin Dow[3], Kirsten J. Lackstrom[3], Maria Carmen Lemos[4] and Rachel E. Riley[5]

[1] *U.S. Department of Commerce, NOAA Climate Program Office, 1315 East West Highway, Room 12212, Silver Spring, MD 20910, USA*

[2] *Western Water Assessment, Environmental Studies Program and Center for Science and Technology Policy Research, Cooperative Institute for Research in Environmental Sciences, University of Colorado, 4001 Discovery Drive, Boulder, CO 80309-0397, USA*

[3] *Carolinas Integrated Sciences and Assessments RISA and Department of Geography, University of South Carolina, 709 Bull Street, Columbia, SC 29208, USA*

[4] *Great Lakes Integrated Sciences and Assessments RISA and School of Natural Resources and Environment, University of Michigan, 430 E. University Ave, Ann Arbor, MI 48109-1115, USA*

[5] *Southern Climate Impacts Planning Program RISA, University of Oklahoma, 120 David L. Boren Blvd., Suite 2900, Norman, OK 73072, USA*

1.1 Introduction

Research on how mankind will adapt to climate variability and change are undeniably important, and yet, traditionally, society tends to turn mainly to physical science for gaining expertise on climate. The Regional Integrated Sciences and Assessments (RISA) program has attempted to remedy this situation by assimilating and generating knowledge that supports the usability of the physical sciences by expanding social and behavioral science on climate and society. We simply cannot understand how best to adapt to climate without gaining knowledge about behavior, policy, institutions, and decision contexts because these aspects often affect the ability of society to respond to and incorporate climate knowledge. Climate research is not only a study of physical processes and impacts, but also a study of individuals, communities, and institutions.

From the beginning, the RISA program has included a human dimensions research element. The number of social scientists in the RISA teams has grown significantly over the course of the program as NOAA staff overseeing

Climate in Context: Science and Society Partnering for Adaptation, First Edition.
Edited by Adam S. Parris, Gregg M. Garfin, Kirstin Dow, Ryan Meyer, and Sarah L. Close.
© 2016 John Wiley & Sons, Ltd. Published 2016 by John Wiley & Sons, Ltd.

the RISA program made deliberate decisions over the years to ensure that social science research was funded over the long term and in a continuous manner. This support allowed the RISA teams to undertake the kind of work discussed in this chapter.

The focus of social science research and the methods used vary from team to team and have evolved and expanded over the 20-year history of the program. In the early days of RISA, understanding the context of decision-makers coping with climate challenges focused mostly on assessing user needs, understanding social and institutional constraints in the use of climate information, and the economic value of forecast information. The network of researchers and range of approaches then grew to incorporate the analysis of risk perception and how decision-makers dealt with uncertain information, assessing the vulnerabilities of different socioeconomic groups to climate, and research on ways to communicate climate information. Over the past decade, the National Research Council (NRC) of the U.S. National Academy of Sciences has called on the federal government to increase its efforts in human dimensions research, build a larger community of researchers focused on these issues as they relate to climate, and use these efforts to build stronger national assessments of climate impacts and adaptation [1].

More recently, RISA work has expanded to include identifying and analyzing information flows across networks of scientists and decision-makers and figuring out how to support these networks by working with key individuals or nodes and providing usable information to them (see Chapter 4). Moreover, as the number of published findings from empirical social science research increased, comparative and meta-analysis studies have emerged. These analyses focus on explaining why users disseminate knowledge across RISAs [2] and how different decision contexts shape what RISAs do across regions to meet different users' needs [3].

Most discussions on the RISA approach tend to highlight the iterative nature of how the researchers interact with decision-makers rather than the methodologies involved (e.g., [4]). To understand the decision context of the planners, managers, and communities with which the teams interact, RISAs draw from a range of social science methods and do so in an interdisciplinary and social–physical science setting.

In this chapter, we discuss the approaches used by four RISA teams to understand the context within which decision-makers operate and use information. Some of the approaches are formal and are based on social science research methods, such as survey and network analysis, and others are more informal based on long-term engagement with stakeholders as well as being present at decision-maker meetings. RISAs learn a great deal about context from both the formal and informal methods. In this chapter, we use the term "decision-makers" to refer to those in the public and private sectors making management and planning decisions. For us, the term "stakeholders" is a broader one that includes other information providers as well as decision-makers and the public.

The four teams from which this chapter's lessons are drawn include Carolinas Integrated Sciences and Assessments (CISA), Great Lakes Integrated Sciences and Assessments (GLISA), Southern Climate Impacts Planning Program (SCIPP), and Western Water Assessment (WWA). Where appropriate, we have drawn examples from other RISA teams as well. The chapter is not meant to encompass all of the social science work undertaken by RISAs, but instead provide some thoughtful insight into the approaches used to understand the context as drawn from the experience of 4 of the 11 RISA teams as well as from a manager of the full RISA program who has observed the breadth and depth of the approaches taken over the years.

1.2   RISA overall approach to ongoing engagement

RISAs are designed to produce new knowledge through fundamental research and to increase the usability of existing knowledge through collaboration with decision-makers and climate information providers. The RISA teams have long and diverse experience with stakeholder engagement of many kinds [5]. RISA researchers regularly and extensively participate in meetings with decision-makers and listen to their concerns, promoting a two-way learning and trust for the knowledge they produce. They also formally study some cases of stakeholder interactions to understand and build a theory on its role in increasing climate information usability.

As Dilling and Lemos [6] observe, iterative engagement between producers and users does not happen in a vacuum and getting it started may take an organization that is willing to foster, and often create from scratch, the conditions necessary to produce usable knowledge. Many decision-making entities produce as well as use knowledge, as do academic researchers thus adding complexity to the analysis of the flow of knowledge or information. Collectively, RISAs have been willing to address this complexity by catalyzing interaction through both formal and informal channels among researchers, decision-makers, and stakeholders.

Coproduction, where new knowledge and the application of that knowledge are produced as a joint venture between scientists and decision-makers, often benefits from interdisciplinary research that draws from the natural, physical, and social sciences as well as interactions within and across research teams. For RISA teams, methods applied to understanding the decision context are often part of, or at least precede, a coproduction process. For example, when severe droughts began to grip the U.S. southern plains in the fall of 2010, SCIPP was able to build on its understanding of the region's drought planning context, and the team quickly identified the need for improving communication with decision-makers about local drought conditions and the strategies that could be employed to manage drought impacts across the region (see Chapter 9). SCIPP collaborated with the National Integrated Drought Information System (NIDIS), a NOAA Regional Climate Services Director, as well as local agencies and organizations to launch a series of webinars, in-person forums, and a state-planning workshop to improve drought impact reporting and to address these needs. Informal surveys of participating decision-makers were conducted throughout the webinar series to provide feedback on content as well as input on future topics most salient to decision-makers concerned about drought.

Many of the informal methodologies focus on maintaining and building relationships and supporting ongoing dialogue about climate-related issues. We think of these as informal methods because, while there is design to the engagement, it focuses mainly on the process rather than on the scientific outcome (e.g., publishable observations or systematic data collection). Instead, it builds the foundation for further partnerships by helping a RISA team gain insight into decision-makers' working context, improve the "information broker" skills of team members [6], and build knowledge networks. Many of these informal approaches resemble participant observation techniques, such as attending annual meetings organized by decision-maker groups and other partners, connecting with decision-makers during breaks, presenting posters at regional and sectoral conferences, supporting community educational events, and serving on various regional committees. Through these various channels, RISAs are listening and noting the issues that concern decision-makers and can offer insights from climate-related sciences to these communities.

RISA teams also bring people together in workshops, meetings, and conference calls, and all teams have regional decision-makers serving on their advisory committees. More recently, there has been an effort to formalize these approaches into quasi-experimental methods without sacrificing their main goal, which was, increasing climate information use in practice. For example, rather than organizing stakeholders directly, GLISA has created an external regional competition for funding boundary organizations—organizations that already bring scientific information to a set of decision-makers. As a result, these local organizations help to bridge clients' needs with GLISA climate information producers. In doing so, GLISA has created boundary organization chains that leverage resources (e.g., financial and human) and spread transaction costs (e.g., trust building and legitimacy). To date, GLISA has funded five of those boundary organizations (and started funding five new ones in 2014) that have agreed to document their interaction and collect baseline and comparable observation data with stakeholders for the duration of their project. While the main goal is to engage stakeholders, the project's comparative design will likely generate valuable data to understand opportunities and strategies to increase information usability [7].

Within their social science focus, RISAs employ formal social science methodologies such as surveys, interviews, and social network analysis for specific research purposes and projects. As the social science capacity within the RISA program has expanded, the range of questions and techniques employed has also grown. These formal methodologies are generally selected to complement the informal efforts described above and vice versa. For example, as GLISA was beginning its engagement in the Great Lakes area, they conducted an analysis of documents produced by stakeholders about their needs, rather than conduct formal surveys or interviews that could contribute to potential stakeholder burnout among potential partners [8].

1.3 Key research questions for understanding context

More than two decades of research and experience demonstrate that usable climate information demands that the process used to create the information must not only be scientifically rigorous but must be acutely aware of the context in which information might be used (see [6] for a recent summary). Research questions related to understanding the context of decision-makers are best posed at the beginning of a RISA endeavor (e.g., the launching of a new RISA or the start of a new project) so that the team enters the process with an understanding of the challenges and opportunities the decision-makers encounter. Moreover, involving decision-makers in the initial framing of those research questions is valuable especially as the team efforts get underway.

Although each RISA investigator and/or team comes up with their own set of questions for understanding the decision-makers' challenges, the questions in Box 1.1 are an illustration of the kinds of questions that can lead to a better understanding of that context. There is usually a large amount of upfront work that needs to be done (see Q1 and Q2 in Box 1.1) before directly asking decision-makers about their needs (see Q4 in Box 1.1).

Box 1.1 Types of questions applied to understanding context.

Q1. What is the existing decision-making context with respect to climate?
   a. What decisions are climate sensitive? How sensitive?
   b. What are the time frames in which climate-sensitive decisions are made?
   After this suite of questions, there is often a decision point as to whether to ask directly about needs (Q4) or to ask about the constraints to using climate or scientific information more broadly (Q2 and Q3). The answer depends on the decision-makers and their context.
Q2. What are the contextual factors that influence decision-making and use of climate information? For example, how do the political, social, and economic environments in which people operate affect their willingness to use climate information?
Q3. What are the intrinsic factors that influence decision-making and the use of climate information? For instance, is climate information accessible and available at appropriate temporal and spatial scales? Do decision-makers consider the information credible, legitimate, and salient?
Q4. What are the specific climate information needs of decision-makers (e.g., resource managers, planners, and communities)?

1.3.1 Analyzing the existing decision making context

As a first step to understanding context, by reviewing the literature, researchers might be able to identify broad categories of sectors, such as agriculture, water resources, and emergency planning, where decisions are climate sensitive. Some initial research drawing from primary and secondary data to look at the major industries, sources of economic activity, unique regional attributes, or populations that are directly affected by shifts in climate can help to narrow down these early suppositions. Being aware of the cultural, social, and political values of key organizations that are climate sensitive is important, but their norms and values can be embedded in decision-making and often become more apparent as trust is built. Thus, a more deep contextual knowledge can emerge over time and with ongoing relationships.

In some cases, decision-makers may initially be seeking assistance in understanding their sector's climate sensitivities. Furthermore, some climate-sensitive sectors may view climate as only a second- or third-order issue because of other economic or social forces in play. SCIPP found this to be the case for agriculture during a survey of needs in Oklahoma; agricultural production is highly sensitive to climate, yet producer profits are tied to financial investments where climate plays a lesser role. Researchers who are not experienced with bringing climate into discussions of decision-making and policy may find it particularly challenging to get involved in these multidimensional discussions.

Another critical step in understanding the decision-maker context is to analyze the time frames in which climate-sensitive decisions are made. For example, some decisions are operational, some occur as part of planning processes, and some are part of making long-term projections. In water management, for example, key decisions such as the release of water from dams are made only at certain times of the year, and to be relevant, information must be available at the appropriate time [9] (see Chapter 2). Decisions to improve or build infrastructure, such as reservoirs or groundwater distribution systems, have long-term implications and must consider climate patterns and their potential shifts over decades into the future. Another example is the importance of the seasonal time frame to the energy sector. Seasonal planning is important to the energy sector, especially during the fall in Oklahoma, because decisions must be made about whether to stock up on extra power poles and wires to improve recovery efforts after devastating ice storms [10]. Further, forest managers in the Carolinas make a range of climate-sensitive decisions at many different time frames. They use hourly forecasts to monitor wildfire conditions, seasonal and annual outlooks to schedule prescribed burns, and long-term datasets when managing timber resources and conservation areas. Managers are beginning to consider how future changes in temperature, precipitation, and the prevalence of extreme events, might impact species selection, hardiness criteria, and biological threats such as invasive species [11]. Developing a sense of the key decisions and associated time frames within an organization is essential.

Understanding the decision context also requires knowing where and how organizations or individuals might already be using climate information. Using methods similar to those discussed above, social science researchers develop specific protocols to learn where decision-makers currently obtain information from, why they use and trust it, and how it helps them accomplish their mission. While new information does not necessarily need to share the same attributes, it is helpful to know about any prior experiences (either positive or negative) using climate information and how those experiences might affect any future attempts to provide usable climate information to those decision-makers.

RISA researchers use multiple methods to assess the decision-making context with respect to climate, including participant observation, interviews, questionnaires, focus groups, and review of the operational documents of organizations. Written documents such as policies, plans, and minutes of council meetings are helpful for gaining a background understanding of an organization's decision context. Stakeholder-institutional analysis is another approach to uncover the broader decision-making context. It involves identifying and examining who makes decisions, the roles and responsibilities of different stakeholders, the rules and organizational arrangements that guide or structure management decisions, what information or knowledge is used, and how risks and uncertainties are perceived and managed. For example, GLISA's "Assessing Assessments" project built a comprehensive database of stakeholder characteristics and needs in the region and identified engaged stakeholders and information about the organizations they belong to (characterized by scale, sector).

RISA researchers work directly with decision-makers such as water utility managers, city planners, or farmers. Meeting directly with decision-makers through a combination of dedicated workshops and interviews, and observations of the decision-makers in their working environment can give the researchers in-depth information about the decision context. In order to develop the Dynamic Drought Index Tool (DDIT) (see Chapter 8) and other drought information and mapping tools, CISA researchers initially worked with decision-makers to understand the drought management context. Methods included formal meetings with state drought response committees and document analysis (i.e., assessing state level response plans and the Federal Energy Regulatory Commission (FERC) licenses that regulate water availability in many of the basins in the Carolinas). Much of the progress the WWA has made in understanding the regional context for water-related decision-making has come through the use of iterative processes that have built an understanding between researchers and decision-makers over several years [12].

Toward this aim, all RISAs have at least one full-time, year-round program manager focusing on the team's overall interactions with decision-makers, and many RISA researchers work with intermediaries (e.g., extension services, NGOs, and other boundary organizations) between the research community and the decision-making communities (see Chapter 4; see also [7]).

1.3.2   Analyzing factors that affect climate-sensitive decisions, including the use of climate information

Understanding both the contextual and intrinsic factors that affect the use of climate information is an important step in producing salient and effective decision support for decision-makers [6].

1.3.2.1   Contextual factors

Developing usable information requires that researchers are aware of and consider how decision-makers operate within a particular regional context, pursue different organizational goals and objectives, and face a variety of political, social, and legal pressures [4,13]. Contextual factors include the organizational and institutional structures that shape decision-making as well as the social, cultural, political, economic, and physical processes that affect the vulnerability of sectors or communities to climate. In addition to understanding the decision process itself, research projects should also take stock of the broader suite of factors that might be increasing or decreasing the sensitivity of the region to climate or the ability of decision-makers in the region to use climate information.

1.3.2.1.1   Organizational and institutional constraints and opportunities:

The options or flexibility that a community or sector faces in dealing with their climate-sensitive decisions can greatly impact their interest in engaging climate information. If the options are limited or severely prescribed, it will be much more difficult for managers to use new information no matter how accurate and credible they are. For example, in many regions, water management decision-making is often legally prescribed, and managers must follow very specific rules in making decisions about water allocation and use. These rules may limit water suppliers in adapting decision processes to incorporate new information. In some cases, only designated types of information can be used (e.g., rule curves that regulate reservoir operations). Nonetheless, there are also many examples of creative and collaborative processes that have been formed to develop legal, robust solutions to new information about climate risk, such as water banking and informal water trading.

Often, the decision space, or "the range of realistic options that can be used to resolve a particular problem," [14, p. 9] varies by management level and spatial scale. For example, a group of "water managers" may include municipal officials, federal agency managers, and state government water departments—all of whom operate in different decision spaces and contexts [13]. Each decision-maker has only a limited area of responsibility and authority to make decisions as well. In research projects, it is critical to understand what role the person participating in the study plays in the decision process and what authority they have to make decisions in order to place their information in the appropriate context.

1.3.2.1.2   Political, social, and economic constraints and opportunities:

Some of the contextual factors that affect the use of information come from outside of the organization itself and include political considerations, different levels of risk tolerance, and costs of available options. Many RISAs have found that engaging in dialogue on climate variability and extremes can be a useful entry point for working with stakeholders that might otherwise be unreceptive to discussing climate change. In some cases, extreme events can help catalyze stronger relationships between stakeholders and researchers as stakeholders seek to cope with responding to or planning for a future event. For example, the drought of 2002 in the WWA region provided a tremendous window of opportunity to open a dialogue about the role of climate variability on water supplies; this type of dialogue was not as pervasive or perhaps even acceptable prior to that point. Interest in regional institutions providing climate information grew after that precipitating event and expanded to an interest in the area of climate change impacts on local water supplies [12].

Political constraints can affect communication and messaging as well. When SCIPP speaks about climate change they focus on climate variability because extremes often occur in the region and have a tremendous impact on communities. In some cases, discussing climate change can carry real risks from a professional standpoint for some resource managers; therefore, sensitivity to those risks is warranted. Likewise, the political environment in the Carolinas, particularly on the state level, is unsupportive of climate change science and initiatives to adapt to climate change. Where adaptation activities are occurring, they are frequently mainstreamed into other types of activities and/or are framed or communicated in alternate terms. CISA research and engagement efforts are sensitive to framing and take into account these constraints on decision-makers [15].

The social and economic tolerance for risks can also differ regionally and affect the willingness of decision-makers to incorporate new information into decision-making and identifying new options. How decision-makers perceive risks and address uncertainties should also be considered and incorporated into decision-support systems. RISA researchers working with agricultural decision-makers and agricultural extension agents may need to consider not only regional and local conditions, but also how international competition and management regimes affect supply and demand. Researchers working with Native American groups will need to understand the historical context and legal frameworks in which these communities make decisions.

As might be surmised from the above discussion, the ability to understand and navigate the contextual factors that affect the ability of an individual or organization to use climate information must be developed through years of interacting, maintaining relationships, and building trust. What makes RISA work unique is the commitment of researchers to develop sustained interactions and in-depth, long-term relationships with stakeholders. Ongoing engagement between scientists and decision-makers contributes to mutual learning and appreciation for each other's needs and constraints, trust building, and the building of networks through which information can be shared.

Such interactions can also help to facilitate efforts to improve understanding of how to manage uncertainties and risks [14,16]. Lemos and Morehouse [4] use the term "iterativity" to describe the process of ongoing interaction between scientists and decision-makers in the Climate Assessment for the Southwest (CLIMAS). Such processes allow researchers and stakeholders to work together to develop tools and information, test those tools, and share feedback for future improvements [14]. In the Carolinas, for example, Carbone *et al.* [17] gathered information through formal and informal interactions, over several years, with water resources stakeholders to identify needs for drought information, as well as to understand the institutional context in which drought response and management decisions are made.

Lowrey *et al.* [12] also discuss the value of long-term, ongoing interactions between decision-makers and scientists within the WWA. The ability of researchers to know the regional context and draw from multiple methods to collect and synthesize information was crucial in guiding their work with Colorado water managers. In addition, informal meetings and interactions are other means through which RISAs can share information, foster social capital, and build capacity for follow-up activities with decision-makers [18]. CISA researchers frequently use such opportunities to keep current on the stresses, as well as opportunities, that influence decision making at multiple levels.

1.3.2.2 Intrinsic factors

In addition to the contextual factors that affect the use of climate information there are several factors affecting use that are more directly related to the information itself and the way it is produced. Decision-makers often have responsibilities for a given geographic region, population center or particular resource. One of the most frequently heard complaints about climate information, for example, is its lack of regional specificity and the lack of skill in the information that is downscaled from climate models. The DDIT for the Carolinas (see Chapter 8) was developed in response to decision-makers who requested drought information at a variety of temporal and spatial scales.

Accessibility of information is also an important component of use. It means that information is readily available and that users can obtain the information from sources such as online databases or personal contacts. Accessibility also means that information is communicated in a way that is understandable. For example, reports that are written in a nontechnical manner, avoid use of scientific jargon, or provide summary information for decision-makers are more accessible.

RISAs work to improve accessibility by improving our understanding of and developing effective methods for communicating climate information.

They apply communications research and "best practice" strategies in their role as information brokers (in preparing newsletters, outlooks, webinars, conferences, etc. for decision-makers). In some instances, RISA investigators have conducted formal research projects to develop and assess these strategies. For example, CISA researchers have used focus groups and web-based surveys from drought and water managers to obtain feedback on graphical representations of probabilistic information and seasonal climate forecasts [19] as well as to assess effective methods for visualizing drought information [20]. The CLIMAS RISA has used newsletters and synthesis reports as boundary objects to inform stakeholders [21].

1.3.2.3   Identifying climate information needs

All too often, the results of scientific research do not inform decisions made by practitioners, despite the intention to produce usable information. We do have evidence, however, that when we understand users' needs we produce more usable science [22]. Thus, understanding the specific climate information needs of different decision-makers is a vital step in providing society with relevant and useful products and services. This step is often left out of the scientific process, which can lead to information that is not useful to or usable by the audience for which it is intended. RISA teams are fully aware of this problem and recognize the importance of understanding what, when, where, and why climate information is needed in particular decision contexts.

Understanding climate information needs drives the development of RISA products and services such as, but not limited to, online climate tools, educational brochures, in-depth reports, and webinars. For instance, in the beginning stages of SCIPP, its team assessed climate information needs through an online survey that was distributed in 2009 to hazard planners across its six-state region [23]. The survey was distributed again in the spring of 2013 to understand what changes occurred since SCIPP was established in 2008. The assessment focused on hazard planning activities across the region and included some questions about information needs. The regional sampling provided a broad representation of the needs of SCIPP's stakeholders and reached decision-makers at various types of agencies and organizations (governmental and nongovernmental) and at various levels (local, state, tribal, federal, etc.). The results then guided SCIPP's research, products, and service efforts.

In contrast, further along in SCIPP's development in 2010–2011, researchers used semistructured interviews to assess climate-related needs in Oklahoma [10] and along the Gulf Coast [24], which are subsets of the SCIPP region. During this stage of SCIPP's evolution as a RISA, the team chose to go more in-depth in order to understand the individual and agency-specific needs, to build relationships that are important for an evolving RISA, and to contribute detailed information about needs to inform the federally-led U.S. National Climate Assessment.

Needs are often complex and may require some conversation. Furthermore, in-person interviews provide the decision-maker the opportunity to begin building a relationship with the interviewer (the information provider or liaison in many cases), which may be vital for decision-makers to feel they have a stake in the issue. In some instances, a need may be identified in an interview that can be met with existing resources. For example, an engineer who participated in the Oklahoma needs assessment asked about historical precipitation data during his interview and a water resource official was interested in soil moisture data. In both cases, the interviewer was able to provide the decision-makers with the data shortly after the meeting.

Workshops are another useful means for determining climate information needs. While the dialogue is typically not as in-depth or specific as during a one-on-one interview, a group discussion can generate ideas that may not otherwise surface. SCIPP employed this method in 2011 when they helped facilitate a workshop with Oklahoma tribal representatives [25]. A workshop setting also fosters relationship building that is often an important component to meeting climate-related needs.

Similarly, WWA has engaged in multiple means of ascertaining user needs, including interviews, workshops, document review, and other informal meetings [12,26]. In some cases, the needs of stakeholders in the context of climate information was not apparent to them until they gained a basic understanding of climate and how it affected the resources they were managing. For example, after a significant drought event, interest at the state level, and the mobilization of new capacity to engage regional stakeholders on preparing for drought (i.e., the formation of NIDIS), WWA held a series of workshops around the state of Colorado called "Dealing with Drought." Participants ranged from water providers and managers, tribes, community leaders, environmental conservation groups, academics, and federal lands managers. The goals included both sharing knowledge on drought and climate variability and change, as well as hearing from stakeholders as to what their concerns were [27]. As a result of the workshops, many stakeholders expressed that they did not feel they had sufficient information about how drought affected the resources they managed. WWA also learned that the stakeholders in the region did not feel sufficiently aware of the informational resources that were available, and that they did not know how to use the information that was available. These workshops were invaluable not only for sharing valuable information with stakeholders, but also for making WWA aware of the needs that were still unmet in the region.

In addition, researchers choose methods to get people to think about their needs and to sensitize them to or develop their thinking about climate and climate information. For instance, some approaches involve, as part of the participatory process, exposing participants to information that may be useful to them. This expands the participants' understanding of their specific needs within the context of science that is currently available as well as could be possibly available in the near term. RISAs frequently generate products that are used and interpreted differently by different groups. Such "boundary objects" can assist both researchers and decision-makers in identifying and developing shared understanding [22,28]. Examples of boundary objects include GIS tools and interactive maps, syntheses reports, climate scenarios, and planning documents. For example, CISA helped to develop the Vulnerability and Consequences Adaptation Planning Scenarios (VCAPS) process as a tool to facilitate dialogue and information sharing between scientists and decision-makers, as well as between decision-makers within a community. This process helps local decision-makers identify climate impacts and needs in their communities and identify and develop response and planning strategies to meet those needs.

1.4 New and evolving area of RISA research: analyzing knowledge networks

In the past few years, scholars focusing on the ability of different systems to respond to climate change impacts have increasingly highlighted the role of knowledge networks as both harbingers of positive normative characteristics (they build trust, amalgamate different kinds of knowledge, and build adaptive capacity and resilience) but also as de facto disseminators of information and innovations [13,29,30].

A network is an "entity consisting of a collection of individuals and the ties among them" [31, p. 5]. Because they map out exchange among actors, network studies are particularly amenable to exploring how information and innovation "travel" or "diffuse" (or not) among different social actors beyond their immediate spatial and social context. In this sense, they can represent a powerful tool to complement place-based analysis of the interaction between producers and users of climate information. They can also help explain patterns of slow diffusion especially concerning preventive innovations such as using climate information for guiding adaptation and for disaster preparedness.

Through interpersonal contacts within networks, decision-makers get acquainted with new ideas, are able to "borrow" from other members' experiences to gauge the compatibility of new tools with their own values and needs, and disseminate the advantages and disadvantages of these new tools to other potential users [32].

Although RISA teams have long been interested in where decision-makers go to access climate information, it is only fairly recently that some of the teams are formally studying the flow of information through networks and the role of knowledge networks in that information flow. In the context of RISAs, formal studies of networks, how they disseminate (or not) climate information or how climate information influences the role of networks in climate-related action have been relatively few (however, see [7,33,34]). Nonetheless, the potential for network studies to inform efforts to enhance climate information usability remains high for RISAs. This is particularly true regarding the influence longer-living RISAs might have had in fostering user networks beyond their originally targeted stakeholder groups. Some of the RISAs have started activities early enough to allow an assessment of how their first clients (early adopters) might have shaped (positively and negatively) knowledge dissemination through the many networks they populate. Moreover, network studies can identify how knowledge of different concerns about climate change impact evolves and amalgamates across different groups of researchers and stakeholders. For example, GLISA is now carrying out two studies that seek to identify how a knowledge network focusing on climate change in general and another on lake levels have evolved in the Great Lakes region. In the first case, the research tries to identify how the climate network has grown in the region since the early 2000s and the different roles organizations and personnel might have played in increasing diversity (across scales and sectors), scope and approaches to respond to climate change impact. The study then includes interviews with network members in the water sector to understand whether and how they use climate information. The second study focuses on meetings, mini conferences, and other venues in which climate change scientists encounter stakeholders and policy-makers with concerns in the Great Lakes region. It uses an analysis of the network structure to select subjects to be interviewed, making sure there are representatives of different types of venues, as well as those who bridged between venues. The study includes a survey of all the subjects focusing on their beliefs about lake levels and freeze–thaw cycles in the Great Lakes with the ultimate goal of learning how information about lake levels and freeze–thaw cycles circulates among scientists and policy-makers. Likewise, the Pacific RISA is currently identifying and mapping the flow of information and communication channels about climate and fresh water resources across the Pacific Islands region of the United States by applying a network survey to a large group of professionals.

The literature on diffusion of innovation identifies three groups of variables by three main categories that affect dissemination: characteristics of innovations, characteristics of innovators, and environmental context [35]. In the RISA-related literature, there is relatively robust empirical analysis focusing on the first two factors (see, e.g., recent literature reviews [36,16]); environmental contexts have been less explored [37], especially regarding the broader social networks where individuals make decisions and information disseminates. By mapping out both the strength of ties and the role of individuals in both disseminating and gate-keeping information, network studies may provide producers of climate information with critical knowledge to strategically target stakeholders and venues (two-mode networks) that can amplify information usability. They can also identify factors that build or undermine trust between personnel and explore the role of policy entrepreneurs to explain policy choice and the dissemination of policy innovation, and identify individuals who "bridge" different clusters thereby potentially accelerating information diffusion and policy-oriented behavior [33]. For example, a study carried out by CISA has found that in politically inhospitable environments, with limited support for explicit climate change activities, ad hoc networks, as well as having a variety of informal opportunities to meet and share information with other decision-makers, are important components of capacity building [38].

1.5   Factors affecting choice of methods

Generally, more engaged methods provide a more in-depth understanding of the decision-makers' concerns and help build long-term relationships, which are important to building and maintaining partnerships. The choice of methods takes into account the research questions, the types of decisions and decision-makers (e.g., farmers, state officials, community planners, federal agency) and the type of relationship researchers have with the decision-makers. It is important to be very clear about the benefits and goals of the research when regional partners are also research participants, especially where partners' actions are being "observed" by RISA researchers.

For many purposes, interviews or other participatory approaches can be used while working with regional partners to identify a smaller group of people with specific expertise, specific needs, or a deep knowledge of a group or issue. Certain research questions benefit from understanding the perspectives of high-level officials whose schedules and personal preferences may only accommodate short, focused telephone or in-person interviews. However, if one has developed a strong understanding of a topic when working

Table 1.1 Pros and cons of _formal_ methods applied to understanding context research questions.

Methods	Pros (i.e., good for which purposes)	Cons (i.e., not as effective for which purposes)
Written questionnaires and surveys (on-line or mail)	Inexpensive (online), broad participation possible (gain perspective from variety of decision makers). Relatively easy way of creating baseline data for evaluation and longitudinal studies.	Does not provide an opportunity to query the specific data or information in which the participant is interested. May not provide enough context to develop appropriate engagement materials (e.g., handout, online tool). Online survey is ineffective when decision maker is not always near a computer (e.g., farmer), prefers interpersonal communication (e.g., tribal representatives) or is less accustomed to working via email/internet. Rates of return can be very low, especially if there is not a previous relationship with potential respondents.
Qualitative interviews (semi-structured or structured)	Promotes relationship building, depth of information, may foster future collaboration. Facilitates an understanding of individual or agency-specific needs. Allows for a two-way dialogue, and the interviewer can ask follow-up questions that clarify the interviewee's response.	Results often not generalizable. Transcribing interviews is time-consuming.
Participatory workshops and focus groups	Interactions can occur more broadly among participants as well as researchers; longer format can build rapport over multiple days. Allow for the progression of awareness to occur so that needs that may not be obvious in the beginning can become known. Similar to interviews, follow-up questions can be asked during a workshop setting, which allows the facilitator to get as specific information as possible, which can lead to meeting a need(s).	Limits number of participants. Requires substantial time commitment of stakeholders. Some groups can be dominated by one person.

(continued overleaf)

Table 1.1 (*continued*)

Methods	Pros (i.e., good for which purposes)	Cons (i.e., not as effective for which purposes)
Document analysis	Stable over time (i.e., can be referred to again and again); unobtrusive (avoid stakeholder fatigue). Can provide background and baseline understanding of decision context and stakeholder concerns.	Documents not always reflective of the full range of opinions; can be biased to certain perspectives; cannot be queried. All relevant documents may not be locatable; needed information may not exist.
Network analysis	Allow for the mapping of networks and understanding of nodes and strength of relationships (sociograms).	Formal network studies can be costly in terms of time and financial resources, especially when using snowballing sampling. Limited generalizability and level of embeddedness in social contexts, which can make exploring causal inferences challenging and necessitating further analysis (quantitative and qualitative) to identify drivers of specific behavior and outcomes [33,39]. Respondents often are reluctant to provide interviewer with names of other members of the network.

Document analysis includes the review and content analysis of documents relevant to the area of study, including documents such as meeting minutes, reports, white papers, operating plans, media reports, and the like. The pros and cons are cited from [40].

with a small group and one needs to understand how it applies to a large number of decision-makers (e.g., county emergency managers or farmers) across a large area, the scale, logistics, and expenses may make a written or e-mailed questionnaire the best approach.

It is important to find a balance between working with decision-makers as partners in the coproduction of knowledge and conducting social science research that involves studying the users of information (i.e., those very decision-makers with whom we are working to produce useful climate information). We also have to be careful to avoid "stakeholder fatigue" (as well as "researcher fatigue"), as these efforts require considerable investments in terms of staff time and resources and commitment from

Table 1.2 Pros and cons of _informal_ methods applied to understanding context research questions.

Methods	Pros (i.e., good for which purposes)	Cons (i.e., not as effective for which purposes)
Participant observation	Promotes understanding of decision contexts and processes, interactions among participants.	Some decision makers and agencies may be uncomfortable being observed; participants are not as forthcoming with information. May not be able to ask questions and/or obtain the information for which you are looking.
Ongoing regional presence/ engagement	Improves effectiveness of outreach efforts. Builds trust with decision making community	Time consuming.
Co-production of research design and analysis	Obtain buy-in from decision makers from the start and throughout the research project. Improves decision maker's knowledge of science and the chances of his/her adoption of new information.	Decision makers do not always have the time, resources, and commitment needed to co-produce knowledge with scientists. Can lead to stakeholder fatigue. Desire of decision makers to use best and worst case scenarios can lead to unlikely projections.

both stakeholders and researchers. Formal surveys, interviews, trainings, and workshops (i.e., sustained and iterative interactions) are critical in supporting the overall process of producing useful information, but they are time-consuming and resource-intensive for all involved. Finding new and perhaps effective remote ways to engage stakeholders can help both to address these issues and to increase the ability of RISAs to reach a larger number of users. [7,22]

Using document analysis, web searches, participant observation (e.g., attending meetings, workshops), and informal interactions to obtain information about decision contexts allows RISA researchers to be sensitive to the time constraints of decision-makers, avoid stakeholder fatigue, and develop and deploy appropriate engagement methods. Moreover, using many sources of data and information allows researchers to integrate information and research findings and ultimately generate a deeper, more robust understanding of how climate information is used, and the political and social context in which decision-makers operate. RISA researchers try to avoid fatigue by building on and using existing efforts, sharing information across networks, and ensuring that there is a clear purpose in mind for stakeholder engagement activities.

For those wishing to pursue various approaches to understanding context, we have compiled Tables 1.1 and 1.2, which summarize the advantages and disadvantages of the different methods discussed earlier.

1.6 Conclusion

RISAs have extensive experience in research and engagement to understand the contexts in which managers, planners, agencies, and communities make decisions with respect to coping with and adapting to climate. A range and mixture of formal and informal research and engagement methods are used to understand these decision-making contexts. These methods are used to understand the existing context—a fundamental first step in effectively engaging those who are affected by climate—as well as potential contexts for information use. These methods are also helpful to determine the specific needs for improved science, services, and products. The mixture of methods often leads to a richer understanding of context as well as contributes to broader social science research on the human–environment interface.

One of the emerging questions for RISA teams is whether everything the team works on needs to be coproduced or whether some products can be developed by just knowing what the stakeholder needs and having an understanding of the context in which the information might be used.

The degree of coproduction needed is a complicated issue and one with which the RISAs are wrestling. Extensive coproduction could potentially contribute to stakeholder fatigue due to the substantial investment of time sometimes needed by stakeholders involved in the research project. The newly evolving area for RISAs of analyzing knowledge networks could shed light on this issue as we improve our understanding of how information flows across different organizations and the roles that these organizations play in producing or coproducing information.

Disclaimer

The scientific results and conclusions, as well as any views or opinions expressed herein, are those of the authors and do not necessarily reflect the views of NOAA or the Department of Commerce.

References

1 National Research Council (2009) *Restructuring Federal Climate Research to Meet the Challenges of Climate Change*. National Academies Press, Washington, DC.

2 Kirchhoff, C.J. (2013) Understanding and enhancing climate information use in water management. *Climatic Change*, **119** (**2**), 1–15.

3 McNie, E.C. (2007) Reconciling the supply of scientific information with use demands: An analysis of the problem and review of the literature. *Environmental Science & Policy*, **10**, 17–38.

4 Lemos, M.C. & Morehouse, B. (2005) The co-production of science and policy in integrated climate assessments. *Global Environmental Change*, **15**, 57–68.

5 Pulwarty, R.S., Simpson, C.F. & Nierenberg, C.R. (2009) The Regional Integrated Sciences and Assessments (RISA) program: crafting effective assessments for the long haul. In: Knight, C.G. & Jäger, J. (eds), *Integrated Regional Assessment of Global Climate Change*. Cambridge University Press, Cambridge, UK, pp. 367–393.

6 Dilling, L. & Lemos, M.C. (2011) Creating usable science: opportunities and constraints for climate knowledge use and their implications for science policy. *Global Environmental Change*, **21**, 680–689.

7 Lemos, M.C., Kirchhoff, C.J., Kalafatis, S.E. *et al.* (2014) Moving climate information off the shelf: Boundary Chains and the role of RISAs as adaptive organizations. *Weather, Climate, and Society*, **6** (**2**), 273–285.

8 Dilling, L., Lackstrom, K., Haywood, B. *et al.* (2014) Building resilience to climate shifts: Implications of stakeholder needs and responses for capacity building in the U.S. *Weather, Climate, and Society*, **7**, 5–17.

9 Ray, A., Garfin, G., Wilder, M. *et al.* (2007) Applications of monsoon research: opportunities to inform decision making and reduce regional vulnerability. *Journal of Climate*, **20**, 1608–1627.

10 Riley, R., Monroe, K., Hocker, J. *et al.* (2012a) *An Assessment of the Climate-Related Needs of Oklahoma Decision Makers*. Southern Climate Impacts Planning Program, Norman, OK. www.southernclimate.org/publications/OK_Climate_Needs_Assessment_Report_Final.pdf.

11 Lackstrom, K., Kettle, N.P., Haywood, B.K. & Dow, K. (2014) Climate-sensitive decisions and time frames: a cross-sectoral analysis of information pathways in the Carolinas. *Weather, Climate, and Society*, **6 (2)**, 238–252.

12 Lowrey, J.L., Ray, A.J. & Webb, R.S. (2009) Factors influencing the use of climate information by Colorado municipal water managers. *Climate Research*, **40**, 103–119.

13 Feldman, D.L. & Ingram, H.M. (2009) Making science useful to decision makers: climate forecasts, water management, and knowledge networks. *Weather, Climate, and Society*, **1 (1)**, 9–21.

14 Jacobs, K., Garfin, G. & Lenart, M. (2005) More than just talk: connecting science and decisionmaking. *Environment*, **47 (9)**, 6–21.

15 Haywood, B.K., Brennan, A., Dow, K. *et al.* (2014) Negotiating a mainstreaming spectrum: climate change response and communication in the carolinas. *Journal of Environmental Policy & Planning*, **16 (1)**, 75–94.

16 Lemos, M.C., Kirchhoff, C. *et al.* (2012) Narrowing the climate information usability gap. *Nature Climate Change*, **2 (11)**, 789–794.

17 Carbone, G.J., Rhee, J., Mizzell, H.P. & Boyles, R. (2008) A regional-scale drought monitoring tool for the carolinas. *Bulletin of the American Meteorological Society*, **89**, 20–28.

18 McNie, E.C. (2013) Delivering climate services: organizational strategies and approaches for producing useful climate-science information. *Weather, Climate, and Society*, **5**, 14–26.

19 Carbone, G.J. & Dow, K. (2005) Water resource management and drought forecasts in South Carolina. *Journal of the American Water Resources Association*, **41 (1)**, 145–155.

20 Dow, K., Murphy, R.L. & Carbone, G.J. (2009) Consideration of user needs and spatial accuracy in drought mapping. *Journal of the American Water Resources Association*, **45 (1)**, 187–197.

21 Guido, Z., Hill, D., Crimmins, M. & Ferguson, D. (2013) Informing decisions with a climate synthesis product: implications for regional climate services. *Weather, Climate & Society*, **5 (1)**, 83–92.

22 Kirchoff, C.J., Lemos, M.C. & Dessai, S. (2013) Actionable knowledge for environmental decision making: broadening the usability of climate science. *Annual Review of Environment and Resources*, **38**, 393–414.

23 Hocker, J. & Carter, L. (2010) *Southern U.S. Regional Hazards and Climate Change Planning Assessment*. Southern Climate Impacts Planning Program, p. 41. Norman. www.southernclimate.org/publications/SCIPP_Hazards_Survey_ Report_Final.pdf.

24 Needham, H. & Carter, L. (2012) *Gulf Coast Climate Information Needs Assessment*. Southern Climate Impacts Planning Program, pp. 20. Baton Rouge. www.southernclimate.org/ publications/Gulf_Coast_Assessment_Final.pdf.

25 Riley, R., Blanchard, P., Bennett, B., *et al.* (2012b) *Oklahoma Inter-Tribal Meeting on Climate Variability and Change*. Southern Climate Impacts Planning Program, pp. 22, Norman, OK. URL www.southernclimate.org/publications/Oklahoma_Intertribal_Climate_Change_ Meeting.pdf.

26 Dilling, L. & Berggren, J. (2014) What do stakeholders need to manage for climate change and variability? A document-based analysis from three mountain states in the Western USA. *Regional Environmental Change*, **15 (4)**, 1–11.

27 Averyt, K., Lukas, J., Alvord, C. *et al.* (2009) *The "Dealing with Drought" Adapting to a Changing Climate Workshops: A Report for the Colorado Water Conservation Board*. Prepared by the University of Colorado, Boulder and the US. National Oceanic and Atmospheric Administration Western Water Assessment. URL http://wwa.colorado.edu/publications/reports/WWA-USFS_WestWatersheds_WorkshopReport_2009.pdf [accessed on 07 September 2015].

28 Star, S.L. & Griesemer, J.R. (1989) Institutional ecology, 'translations' and boundary objects: amateurs and professionals in Berkeley's Museum of vertebrate zoology, 1907–39. *Social Studies of Science*, **19 (3)**, 387–420.

29 Olsson, P., Folke, C. *et al.* (2004) Adaptive comanagement for building resilience in social–ecological systems. *Environmental Management,* **34** (**1**), 75–90.

30 Lemos, M.C. (2008) What influences innovation adoption by water managers? Climate information use in Brazil and the US. *Journal of the American Water Resources Association (JAWRA),* **44** (**6**), 1388–1396.

31 Wasserman, S. & Faust, K. (1994) *Social network analysis: Methods and applications.* Cambridge University Press, Cambridge, UK.

32 Valente, T.W. & Rogers, E.M. (1995) The origins and development of the diffusion of innovations paradigm as an example of scientific growth. *Science Communication,* **16** (**3**), 242–273.

33 Frank, K., Chen, I. *et al.* (2012) Network location and policy-oriented behavior: an analysis of two-mode networks of co-authored documents concerning climate change in the Great Lakes Region. *Policy Studies Journal,* **40** (**3**), 492–515.

34 Crona, B.I. & Parker, J.N. (2011) Network determinants of knowledge utilization: preliminary lessons from a boundary organization. *Science Communication,* **33** (**4**), 448–471.

35 Wejnert, B. (2002) Integrating models of diffusion of innovations: a conceptual framework. *Annual Review of Sociology,* **28**, 297–326.

36 McNie, E. (2008) *Co-Producing Useful Climate Science for Policy: Lessons from the RISA Program.* University of Colorado, Boulder, CO.

37 Bolson, J., Martinez, C. *et al.* (2013) Climate information use among southeast US water managers: beyond barriers and toward opportunities. *Regional Environmental Change,* **13** (**1**), 1–11.

38 Dow, K., Haywood, B.K., Kettle, N.P. & Lackstrom, K. (2013) The role of ad hoc networks in supporting climate change adaptation: a case study from the Southeastern United States. *Regional Environmental Change,* **13**, 1235–1244.

39 Berry, F.S., Brower, R.S., Choi, S.O. *et al.* (2004) Three traditions of network research: what the public management research agenda can learn from other research communities. *Public Administration Review,* **64** (**5**), 539–552.

40 Yin, R.K. (2003) *Case study Research: Design and Methods,* 3rd edn. Sage Publications, Thousand Oaks, CA.

CHAPTER 2

Understanding the user context: decision calendars as frameworks for linking climate to policy, planning, and decision-making

Andrea J. Ray and Robert S. Webb

Physical Sciences Division, NOAA Earth System Research Laboratory, 325 Broadway, R/PSD1, Boulder, CO 80305, USA

2.1 Introduction

Advances in climate research over the past few decades have led to significant improvements in the understanding, attribution, and prediction of climate variability. In parallel, advancements in interdisciplinary research have examined if and how subseasonal to interannual climate information has been, or could be, used and the value of this information in policy, planning, and decision-making in agriculture, energy, water management, and other sectors, to manage risks and mitigate impacts [1]. Within these sectors, research has identified cultural and institutional barriers that have limited the use of subseasonal to interannual climate information in resource management and operations [2]. Specific barriers relative to decision needs include the following: that climate information relative to the decision needs is not available at the right time [2,3], and that time scales are not meaningful to potential users [4], do not meet their temporal needs [5], or are not at the appropriate spatial scales [2]. *Useful* as defined by scientists' assumptions of what users need, has been distinguished from *usable* as defined by users' perceptions of their needs [5], by the user's context [6], and produced to directly contribute to the design of a policy or solution of a problem [7].

Thus, a major factor in understanding and clarifying users' needs is dependent on understanding the user context. Scientists must better understand contextual factors such as the organizational and institutional setting, and decision rules [6] in which information is used in order to develop usable information. Sustained interactions and deliberate engagement have been found to be essential to understand the user context, and, based on that understanding, to identify user needs for decision-making, planning, and policy processes, ([7] and Chapter 1). A coproduction model [6] in which scientists and potential users iterate about the state of understanding of climate variability, predictive skill, and potential users' needs is found to be more successful in understanding users' needs. This sustained interaction needs to be iterative among three groups: between users and climate information providers, researchers and providers, and researchers and users [8]. Engagement strategies include the participatory and user-centric approaches used by boundary organizations [9,10] and knowledge brokering [11], and others discussed in Chapter 1. While this deliberate engagement is crucial, it is also time and labor-intensive, and there is little guidance on how to organize and analyze the information collected in order to understand user contexts. Analytical frameworks for organizing information about the user context, and from that, refining user needs, are not well documented (see Chapter 1 for RISA approaches to this challenge).

Informed by over a decade of work, this chapter describes decision calendars as a practical framework to organize information gathered from a variety of engagement and methods. By organizing information about a user context, decision calendars can help overcome constraints and barriers that are relative to incorporating climate information in the decision-making process by identifying entry points and opportunities that influence the use of climate forecasts, for example, the timing when products are most likely to be used by resource managers and their stakeholders. Thus, the decision calendar can also be thought of as a boundary tool (see Chapter 5), intended to move past first-order questions regarding if and how stakeholders use climate information, to more sophisticated and contextual second-order questions to assess what is needed for the information to be usable. Furthermore, beyond identifying usable information, the decision calendar framework identifies entry points in the decision process for usable and actionable information. In this chapter, we define *actionable information* as a science-based knowledge, which is transformed to be readily understandable and immediately available to be incorporated into decision-making.

The introduction to this volume raises several challenges to understanding user needs that the decision-calendar framework can address. Decision calendars provide an efficient way to organize and refine the climate science needs of users as described in the presented case studies of reservoir and fire management. Decision calendars also help users or researchers discern differences between initially "perceived" and "actual" needs for how people could use or take action with science information in the decision process. In so doing, a decision calendar can help decide where to invest resources in the next stage of research or product development. Furthermore, with an expanding array of sectors and potential users seeking climate information, decision calendars can organize the needs of multiple groups to identify common needs, as described below in the monsoon example. Finally, the process of developing decision calendars also provides a space to foster engagement and communication among researchers, potential users, and climate information providers, to provide feedback to climate scientists on usable information products, and to guide use-inspired research. Thus, the value of decision calendars is twofold: the calendar as a research product itself, which illuminates user needs; and the process of development, which fosters engagement.

Our decision calendar approach builds on several analytical frameworks that each contribute to the analysis of the user context, including the critical water problems approach [12], policy sciences [13], and institutional dynamics [14]. These approaches are problem focused [15], versus a theoretical inquiry, recognizing that no explanation covers every case. By developing an understanding of the user's context, perspectives, and time frames it becomes possible to develop or coproduce usable information to address these problems. These frameworks were selected because they address key considerations in climate-related decisions for a particular user context, that is, the users' critical issues and problem focus; their planning horizons and goals; basis for decisions, and, as described in Jacobs and Pulwarty [16], differences in perspectives of different stakeholders as well as between scientists and stakeholders.

Many analytical frameworks lack a temporal dimension of use of products and information. Adding the temporal analysis—a distinguishing feature of decision calendars—to insights from other social science tools provides a way to organize annual and other recurring cycles of decision-making, to understand the role and influence of the longer-term planning cycles, and to evaluate the decision-making process in a climate context [17]. Our decision calendar approach allows identification of critical time periods (so-called "entry points") and formats to provide climate forecasts and information, as well as to infer climate information that is potentially useful, but not currently used. In short, a decision calendar is a framework for organizing information about a user context and related climate knowledge. This information may be collected using a variety of formal and informal methods, including those discussed in Chapter 1. In turn, the decision calendar has informed the work by NOAA RISA programs with stakeholders by guiding climate research priorities and pragmatic decisions on where to focus engagement efforts.

We present three cases of RISA work where decision calendars were created, and also examine the role of decision calendars in developing usable and actionable information. Diverging from the sector orientation common to many studies, these cases focus on specific climate-sensitive societal problems and assess how climate information may inform management responses to address the problem. These projects by the Western Water Assessment (WWA) and the Climate Assessment of the Southwest (CLIMAS) RISAs linked climate information and decision-making. Decision contexts include: (1) reservoir management, from annual to longer-range operations, situated in larger societal and environmental problems that the water systems serve; (2) fire management, and (3) scoping applications of a research program on the North American Monsoon. We then describe several steps in the development of decision calendars. We conclude with a discussion of the contributions of the decision calendar framework, which we believe can provide practical insight into understanding user needs in many contexts.

2.2   Reservoir management

The WWA, one of the early RISA programs (described in more detail in Chapter 11), began in the late 1990s soon after a new generation of dynamical forecasts of El Niño-Southern Oscillation (ENSO) and seasonal climate outlooks were being issued by the NOAA Climate Prediction Center. An early goal of the WWA was to understand how these predictions could be used in water management in the Colorado Basin and the state of Colorado, including its Front Range water providers. However, understanding the climate needs of water managers was in its infancy. We anticipated constraints and barriers similar to those identified in the then-recently-published paper by Pulwarty and Redmond [2]. We needed to understand the particular context of the barriers and constraints for this region, such as water laws and institutions, in order to move on to entry points and opportunities, including who might use the new information and in what decisions.

As a relatively new program, an early task was to narrow the many potential partners in water management and the many potential water problems to address, that is, to make pragmatic choices on who to work with and where to invest time and other resources. A critical water problems approach [12] proved useful to identify topics of most concern to water managers, which were also sensitive to climate. Critical water management problems included changing reservoir operating plans to provide flows for environmental purposes; implementation of the Salinity Control Act; equity issues such as Indian water rights; and competition among uses, such as transbasin water diversions and new uses of water for recreation or aesthetics [17].

Climate variability and change appeared to be most relevant to reservoir operating plans, although climate variability was of more interest at the time.

The primary decision-makers in this case were the U.S. Bureau of Reclamation, Denver Water, and the Colorado River Water Conservancy District. Each management organization was both the decision-maker for its own reservoirs and a stakeholder in management of the basin. But the larger decision context includes who is involved or influences the decisions and what is important to them [13,21], that is, the multiple agencies with "stakes" in the management of the reservoirs and water releases upstream of the critical habitat for endangered species near Grand Junction on the Colorado River. The "stakeholders" in reservoir management include the U.S. Fish and Wildlife Service with authority under the Endangered Species Act (ESA); Xcel Energy, which holds a large, senior water right for hydropower generation; the irrigation districts downstream in the Grand Valley; and recreational uses then without formal water rights such as trout fishing and rafting. Furthermore, two external federal environmental mandates were challenging the reservoir managers to provide water for environmental purposes [18]: the ESA requirement to provide water from reservoirs to assist with habitat restoration and recovery of endangered native fish, and a Federal reserve right for flows for the Black Canyon of the Gunnison (now a national park).

According to the institutional dynamics framework [14], when some fundamental change in policy and/or management has or is likely to happen, the system may be in a release and reorganization phase, and more open to consider new ideas or new technologies such as climate information, and more likely to be open to interacting as partners. Reservoir managers were anticipating that the flow recommendations for endangered fish, under development for several years, would be finalized and changes to operations required to meet the recommendations would be made. These looming requirements provided urgency, or criticality [20] to expand management to meet the new water needs. In fact, reservoir managers had already agreed to try new operating strategies: under an agreement called Coordinated Reservoir Operations (CROS), reservoir managers agreed to evaluate each spring whether if their reservoir supply included flows that might otherwise be spilled that spring and, if so, that year they would voluntarily to coordinate releases with the timing of the peak flows at the critical habitat reach of the river upstream of Grand Junction on the main stem of the Colorado River [17].

From initial conversations, we knew that reservoir operators were considering seasonal climate outlooks, but were not using them in formal operating models. Thus, they knew about the forecasts, but had only progressed to the first stage in a process described by Rogers [21] through which individuals or organizations decide to adopt an innovation. They had passed Rogers' knowledge stage, and had formed an initial favorable opinion that the forecasts might be useful (persuasion stage). A workshop in October 1999 (also described in Chapter 11) was held to provide a forum to explore potential uses and to encourage them to move into the decision and implementation stages, and to engage in activities where they might adopt or reject the seasonal outlooks as an innovation.

At the 1999 workshop, an engineer from Denver Water sketched their decision process and challenge (Figure 2.1). Reservoir managers seek to fill the reservoir by the end of the April–July runoff season, by balancing two competing goals of maintaining some storage space for potential heavy runoff events to avoid floods, but avoid "spilling" water that might have been stored. The CROS program added another goal for the managers: intentionally saving water that might have been spilled to be released or "bypassed" through the reservoirs in a coordinated way to create higher flows in critical habitat reaches of the river. Reservoirs farthest from the critical habitat would release water ahead of those closest to the critical habitat, so that the bypassed flows reach the target area at the same time. This scheduling highlighted the need for accurate flow forecasts made by the NOAA Colorado Basin River Forecast Center (CBRFC), for the volume forecasts for the reservoirs, as well as for the flow levels in the target reaches. Both of these

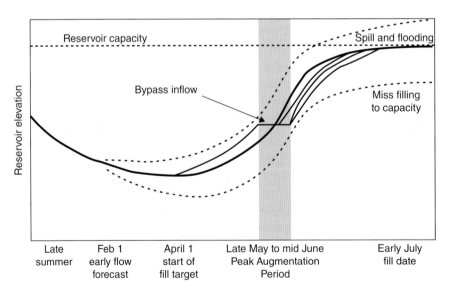

Figure 2.1 Conceptual reservoir hydrograph. Key points in managing inflows into reservoirs, based on engagement with reservoir managers at a workshop and subsequent discussions.

river outlook products were already in use by these water managers, but they thought that other climate predictions, if incorporated into the CBRFC products, might give them an outlook of the wet or dry conditions, and thus more time and flexibility to prepare and manage options.

The workshop participants then outlined what products might be needed and when in their annual decision process [17]. NOAA long-lead precipitation and temperature forecasts in the fall could improve the outlooks of winter snowpack accumulation and estimates for subsequent April–June runoff. Through fall and winter, an understanding of the influences of ENSO on the seasonal evolution of snowpack could lead to more accurate estimates of runoff and the "start of fill" target reservoir level. Throughout the spring, a one-to-two week precipitation and temperature forecast could provide improved estimates of volume and timing of spring peak flows needed to plan peak flows for habitat restoration as well as to enhance flood mitigation operations. By late spring, the NOAA long-lead forecasts could improve demand forecasts for summer irrigation and other water demands. Throughout the summer, 1–2 week precipitation and temperature forecasts could improve estimates of releases for both hydropower generation and, in order to schedule irrigation, mitigation of low flows in the river.

Subsequent to the workshop, a decision calendar (Figure 2.2) was constructed to organize these needs and identify other potential uses, indicating the timing of select planning processes, operational issues, and the timing of potential uses of several types of climate and weather forecasts that could be used to inform planning and operational concerns. The resulting decision calendar identified climate information needs in the reservoir management context and identified other potential uses. The evolution of the decision process was documented throughout the water year (and longer timescales) to identify the timing of needs as well as the potential entry points for climate information.

Another result of this work with reservoir managers was to broaden our conception of "use" of climate information beyond simply incorporating a prediction into an operational or planning model. In our reservoir management studies, we have observed four types of use: (1) Consult: the product is looked up on a web page or received from other source; (2) Consider: after consulting the product, the information is integrated into management deliberations as a factor that could potentially influence decisions, but was not used in operational models; (3) Incorporate: the forecast is assimilated into an operational model that is used in operational decisions; and (4) Dialogue about risks: the forecast is used to communicate with other managers and stakeholders about the risk of certain conditions, the need to take actions, or to justify actions [17].

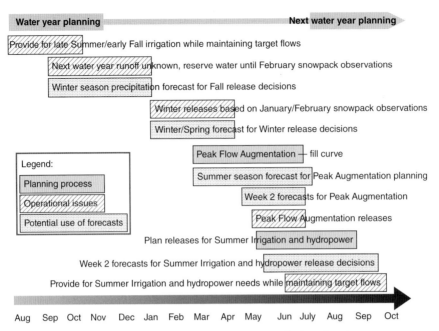

Figure 2.2 Reservoir management decision calendar. Timing of select planning processes (gray bars), and operational issues (dotted bars), for Upper Colorado River reservoirs. Stippled bars indicate the timing of potential use of several types of weather and climate outlooks to address these planning and operational concerns. The width and position of the bars indicates the relevant time periods. For example, in the late winter, improved forecast of the runoff volume (after [17]).

Understanding the ways climate information is used into the decision-making process (consult, consider, incorporate, or dialogue) can provide insights into both its value and limitations. For example, predictions that are not in an appropriate form or are thought to lack an acceptable level of skill for use in an operational model (type 3) may still be used subjectively (type 2) in mental models via the judgment and experience of water managers as described by Jacobs and Pulwarty [16]. It may also still be used in a dialogue about risk (type 4) to convey the climate risks behind a management action intended to minimize impacts while gaining the agreement or support of management or stakeholders. In types 2 and 4, the use of the information may be qualitative or subjective even when information is quantitative [16]. Consulting the information (type 1) is a kind of use. For example, a forecast may be consulted for situational awareness, but no other use may be needed if it raises no concerns that need to be followed up on such as, if drought or high flows are unlikely. On the other hand, a user may consult the information and not use it further because it is not usable for some reason. As part of a decision calendar analysis, the types of use may show different entry points and opportunities for a particular information product. In addition, the decision calendar framework can identify entry points and opportunities for information that is not currently available, or not available as a usable product.

As an outcome of the 1999 workshop and the decision calendar, WWA and its NOAA partners developed ongoing engagement with Reclamation and its stakeholders, which included presentations and participation in water managers' own management meetings and the related Colorado Drought Task Force, WWA-hosted meetings. As described in Chapter 11 WWA's early efforts as a boundary organization were informed and shaped by this early work with reservoir managers and included efforts intended to help water managers understand the available seasonal to interannual forecast information. The efforts also targeted the stakeholders in reservoir management, so they would also understand the potential use of predictions and forecast information and the rationale behind the resulting management actions. WWA and PSD conducted applied research targeted at improving the understanding of climate at key points; for example, work to connect longer-range (8–14 days) weather outlooks to the CBRFC river products [22] and to understand the impacts of multidecadal variability on regional water resources [23].

2.3   Wildfire management

The CLIMAS RISA focuses on assessing climate variability and longer-term climate change in terms of impacts on human and natural systems in the Southwestern United States (http://www.climas.arizona.edu). In early 2000, CLIMAS identified wildfire as a critical management problem and fire managers as ready and willing potential user groups for engagement. The combined impacts of the challenges of increases in acreage burned, growing costs, and greater variations in the magnitude and severity of wildfire seasons in the western United States was making wildfire management an increasingly complicated and resource-intensive endeavor. CLIMAS embarked on workshops and discussions with natural resource and fire decision managers to explore what climate information was needed to manage wildfire, and when and where the information was needed (see Chapter 7). Their engagement process included training to ensure that the basic knowledge and skills needed for interpreting and using climate information were available [24]. CLIMAS's engagement with the wildfire community reaffirmed the strong link between fire danger in the western United States and winter precipitation, the need for a change from reactive to proactive management strategies, various planning horizons for preparedness, and improvement in the use of extended weather forecasts and climate outlooks by agencies to strategically plan wildfire resource allocation and fire use opportunities [24].

To better understand the decision-maker's context, Corringham *et al.* [25] surveyed wildfire management officers and decision-makers in 2002–2003. They asked different individuals and groups to construct annual decision calendars identifying when their own fire prevention and suppression decisions were made, how/if climate information supported these decisions, what information was used, the sources of information, and what other information could have been useful. Survey information was combined with local, regional, and national inputs from the primary federal wildfire management agencies, U.S. Forest Service (USFS) and National Park Service (NPS), and insights from the perspective of broader interagency coordination. Four primary types of wildfire management decisions were common across three states (California, Arizona, and New Mexico): fire suppression, prescribed fire and fuel management, seasonal staffing, and budgeting. Specific decisions involved the Santa Ana winds in southern California, presuppression pile burning in northern California, and onset of monsoon rains in Arizona and New Mexico. Institutional barriers identified included constraints in retargeting funds within a 2-year budget cycle, inflexible authorizing legislation, organizational inertia, temporal mismatches between decisions and forecasts, and risk aversion. These barriers were obstacles in using climate forecasts and other information to better manage wildfire risk.

A wildfire decision calendar was created based on the information from surveys and interviews, and the annual cycle of wildfire decision-making by USFS and NPS management units in California, Arizona, and New Mexico [25]. The sequencing of decisions by region is summarized in Figure 2.3. Weather/climate information for managing wildfire ranged from historic data to provide perspective, to real-time information and short-term spot forecasts and 1- to 5-day-lead forecasts of weather conditions to support tactical decisions, to seasonal and annual outlooks to long-term climate products to support preparedness and strategic planning decisions. As the fire season progresses, outlook and assessments can be updated with shorter-term forecast products while year-to-date conditions are compiled and compared with historical climate averages to determine when more active monitoring and management activities are required.

The decision calendar framework was used to identify entry points for improved use of climate information to guide wildfire management decision-making. The analysis by Corringham *et al.* [25] indicated that

Southern California

Decisions	Jan	Feb	Mar	Apr	May	June	July	Aug	Sept	Oct	Nov	Dec
Suppression						███	███	███	███	███	███	
Rx and fire use	███	███	███	███	███	███						
Season staffing					███	███	███	███	███	███		
Budgeting			███	███								
Special: Santa Ana	███								███	███	███	███

Northern California

Decisions	Jan	Feb	Mar	Apr	May	June	July	Aug	Sept	Oct	Nov	Dec
Suppression							███	███	███			
Rx and fire use				███	███	███	███	███	███			
Season staffing				███	███	███	███	███	███			
Budgeting			███	███								
Special: pile burning	███	███	███							███	███	███

ARIZONA AND NEW MEXICO

Decisions	Jan	Feb	Mar	Apr	May	June	July	Aug	Sept	Oct	Nov	Dec
Suppression			███	███	███	███	███	███				
Rx and fire use			███	███	███		███	███	███			
Season staffing				███	███	███	███	███	███			
Budgeting			███	███								
Special: monsoon						███	███	███				

Figure 2.3 Regional aggregated wildland fire management decision calendars. Months during the annual cycle when information is needed for each of the decisions are indicated by ███ (after [25]).

reliable wintertime La Niña outlooks prior to a fire season would allow decision-makers to adjust budget allocations to manage risks. An improved prediction of the probability of dry lightning strikes prior to summer rains and the timing and magnitude of monsoon rains were highlighted as important for resource allocations. At shorter timescales, 8-14 day forecasts were found to be useful for planning prescribed burn activities. They also identified opportunities for extended weather forecasts and short-term climate outlooks that were critical for fire suppression decision-making and to identify opportunities for fire use to reduce fuel loads and for landscape restoration. Evaluation and communication of weather forecast and climate outlook skill, reliability, and uncertainty in conjunction with wildfire management actions and outcomes would improve decision-makers' understanding of strengths and challenges of using this information to manage wildfire risk. Even as advances in climate information are combined with improved understanding of the decision-making processes, the combination of multiple objectives and differing priorities challenges the use of climate information in wildland fire management practices.

As a result of the engagement and the information needs identified using the decision calendar framework, an ongoing national seasonal fire assessment process has been institutionalized as a partnership among CLIMAS, the National Interagency Fire Coordination Center's Predictive Services, and the Program for Climate, Ecosystem and Fire Applications at the Desert Research Institute in Reno, Nevada with involvement by WWA [26] (see Chapter 7). In recent years, the scope of these regional efforts has expanded to cover all of North America and to provide monthly updates during the wildfire season [27]. These assessments of significant wildland fire potential are used by decision-makers for proactive wildland fire management to better protect lives and property, to reduce firefighting costs, and to improve firefighting efficiency.

2.4   Applications of monsoon research

In the early 2000s, the NOAA-funded North American Monsoon Experiment (NAME), prompted interest on how improved understanding and forecasts of the monsoon might be used, and how to target potential users of the information about this phenomenon. In this case, climate scientists had identified the monsoon as the target of a multiyear process study designed to improve understanding and simulation of the monsoon in coupled climate models in order to predict monsoon features such as onset and decline months to seasons in advance [28]. Around the same time, the National Research Council had recommended that NOAA should "ensure a strong and healthy transition of U.S. research accomplishments into predictive capabilities that serve the nation," [29], and NOAA also was actively seeking strategies to transition research results to operational products and applications provided by the agency [30]. Although the impacts of the monsoon were recognized as being societally important, there was no clear understanding of what was needed by potential users for the results of the NAME program to be used and transitioned to operational products and applications.

A team of CLIMAS and NOAA social and physical scientists assembled to assess this question of how NAME program results could inform users, that is, who might use the knowledge produced by the program, what kinds of decisions would be informed, and when those decisions would need to be informed. They also hoped to provide feedback on priorities to the research planning process for NAME and related programs like the NOAA Climate Test Bed. The CLIMAS RISA was already engaging with stakeholders, studying potential uses of predictions of seasonal climate variability including the monsoon, as well as societal vulnerability to ENSO-driven variability, and drivers of fire risk (discussed in the case above). CLIMAS and NOAA had cosponsored workshops in 2001 and 2006 on monsoon applications that engaged potential users with climate and social scientists [31].

This case of developing a decision calendar differs from the two described above because it focused on the monsoon, which was a more societally relevant weather and climate event, rather than starting with a management problem. Furthermore, rather than developing new engagement, the process built on knowledge of users from existing long-term engagement with stakeholders by the CLIMAS RISA and from the two monsoon-applications workshops. The findings of those engagement efforts were integrated with published studies documenting climate-related stakeholder exposures, sensitivities, and adaptive capacities (see discussion in [31]), as the basis for analysis of needs crossing several different applications sectors.

The multidisciplinary team reviewed and synthesized the available studies and identified five distinct planning and decision applications likely to be sensitive to the monsoon, and their climate- and monsoon-sensitive critical management problems [31]. The five applications likely to benefit from better monsoon outlooks and within-season information were: natural hazards (fire, flood, drought), agriculture in Sonora, ranching in Arizona, urban water management in Arizona and northwest Mexico, and public health. The team identified the key decision-makers, their monsoon information needs, and the potential use of that information (see Table 1 in [31]). A decision calendar framework was used to organize and synthesize the cross-application spectrum of climate sensitivities and vulnerabilities and to identify the timing and common needs or uses of improved understanding of the North American Monsoon (see Figure 3 in [31]).

The monsoon applications decision calendar was intended to identify and guide efforts of future scientists to link user needs for monsoon information to monsoon forecasts and information products. By identifying common, or shared, needs across several application sectors (water, fire, ranching, agriculture, health), this study provided guidance to NOAA on priorities for research and development to advance operational delivery of the needed monsoon information. Ray *et al.* [31] concluded by recommending a regularly issued product focused on the monsoon, and engagement activities to build capacity to use monsoon information, through approaches such as experimental climate products and services that introduce potential users to actionable information. As an outcome of this analysis, CLIMAS and its partners began conducting monsoon outlook webinars and an online Monsoon Tracker for many of these applications groups, as a way to continue engagement, and the monsoon is a regular feature in their Climate Summary, another CLIMAS engagement tool. The NOAA Climate Prediction Center developed a North American Monsoon page with monsoon metrics as part of its suite of their experimental products. The process of developing a decision calendar for the southwest Monsoon identified specific applications areas and management questions for further study and engagement, as well as provided feedback to guide NAME research development, and the development of use-inspired products.

2.5 Developing decision calendars

Developing decision calendars requires, at a minimum, documenting who the decision-makers and their key stakeholders are, key decisions and the timing of those decisions, and their needs as they describe them. In summary, decision calendars provide an analytical framework for organizing information about a user context, including timing of decisions and climate information needs, and then identifying entry points and opportunities for use of climate information. The contextual information may be collected using a variety of social science methods (see also Chapter 1). Developed from experience over more than a decade, the approach described here has been a key tool for RISAs to address second-order questions, that is, moving beyond identifying barriers to understanding how to get beyond or through these barriers.

The use of decision calendars to develop usable and actionable climate information is based on the premise that knowing the issues confronting users (challenges, current use of information, decision process, operations, and planning processes) can guide the development of information products that better inform decision-making, and also inform use-inspired research questions and the co-production of knowledge. Furthermore, climate products desired by a spectrum of users can be identified through collecting and integrating contextual knowledge about different users on what kind of climate information is needed, when, and how, and by conducting needs assessments while recognizing that climate needs are often embedded in decision contexts not exclusively related to climate. No explanation covers every case [32], but by creating decision calendars aggregated across distinct management groups (as in the fire management case), or across multiple applications (as in the monsoon case), common needs for information can be identified.

Methodologies used in developing decision calendars are primarily qualitative and context-sensitive [32], and triangulation among observations from different methods may be used to build confidence in the findings [33]. In contrast to methods that are intentionally detached and observing, engagement and iteration are intentional and crucial in the process of developing a decision calendar. Surveys may be used, but context-sensitive methods are likely to be required to gather more detailed information on the user context (or reanalyze it from prior studies, as in the monsoon case). These methods include interviews with open-ended questions, text analysis, and participant-observation.

Ongoing engagement, or iteration, with users is a critical part of developing decision calendars in order to refine, and over time, update conclusions about information uses. While creating ongoing engagement is not typical practice in many types of social science research, engagement—especially as participant observation—it is a recognized practice in anthropology, along with acknowledging the fact that the system studied is changed by the participation of researchers [34]. Workshops and collaborative projects with decision-makers, participant-observation, and ongoing communication through webinars and other communication tools are all strategic methods to develop and maintain engagement. The development of decision calendars can also be a collaborative mechanism for climate scientists, information providers, and decision-makers to interact iteratively and learn from each other in the co-production of knowledge. The steps in developing a decision calendar are discussed in the following text.

2.5.1 Document decision-makers and their key stakeholders

For a given critical management problem, identifying both the decision-makers and their key stakeholders is required. Methods include analyses of records of past decision processes and other documents (such as minutes of management meetings), public documents or webpages, or prior studies related to a case. Even in a new context or working with a new user group, this analysis can quickly identify problems that are most likely to motivate decision-makers to work with climate scientists. Identifying the key stakeholders who influence decisions is important because climate needs are often embedded in decision contexts not exclusively related to climate and are often related to concerns outside the direct control of the primary decision-maker. For example, in the reservoir management case, although the Bureau of Reclamation had formal authority to operate several reservoirs, they had to do so in a complicated milieu of other concerns, including the recovery of endangered fish and local interests for recreation and trout fishing in the river.

This step may begin with existing documents to develop a general sense of the context, then use in-depth and context-sensitive methods such as interviews or participant-observation at meetings, or through a workshop convened by the RISA, to confirm initial findings and to give a more detailed picture of the decision-maker context. This context may give insights to the reasons behind the use of information, and can assess the potential of that decision environment to take on new information.

2.5.2   Document key decisions, needs expressed for climate, and the timing

This step involves collection of detailed and contextual information about specific decisions and climate information that potentially support decision-making. It builds on the information collected in developing the decision-making context, and as in the first step, context-sensitive methods are needed, such as in-depth interviews with the decision-makers and their stakeholder organizations and others involved in the problem, participant-observation, or engagement through workshops. Information is also collected about the timing of planning and decision processes of key stakeholders because their concerns may significantly affect the primary decision-maker.

2.5.3   Organize information into a decision calendar

In collaboration with the decision-makers and their stakeholders, documenting the evolution of the decision process over time (such as the water year or annual planning), and at longer timescales if appropriate. Engagement is crucial in this step and may occur in a workshop, in conjunction with meetings organized by the users for their own purposes, or in repeated meetings with one or more of the decision-makers and stakeholders. In some cases, the engagement may have occurred during prior studies, and in such cases, the results are re-analyzed. The resulting decision calendars identify critical time periods for climate forecasts and information, and from this analysis, researchers can infer climate information that is potentially useful, but not currently used.

In this step, climate scientists and providers contribute their understanding of potentially predictable aspects of climate, and engage in a dialogue about what climate knowledge might inform the user's deliberations and decisions. The decision calendar framework can be used to document: (1) when and for what decisions and planning processes climate predictions were currently used in any way, (2) needs that have been expressed, and (3) whether the information needed was available or not. Analysis of the decision calendar identifies entry points for climate information that might be used, that is, when and for what decisions and planning processes that climate predictions may be useful. Critical time periods for climate forecasts and information can be identified, and climate information that is potentially useful, but not currently used, may be suggested. The decision calendar then provides an understanding of recurring (such as annual) cycles of decision-making and the longer-term planning cycles they are embedded within.

2.5.4 Continue engagement and iterate to confirm and refine initial findings

The process of developing decision calendars has been a mechanism for creating and sustaining ongoing engagement. In this step, climate scientists and users may identify potential collaborative projects or experimental climate services, such as testing improvements to existing products or new desired products. These activities help to create and sustain ongoing engagement, while also fostering a deeper understanding of needs and two-way learning based on that engagement, thus satisfying two important broad goals of the RISA program.

2.6   Discussion and contribution of this framework

This chapter illustrates how decision calendars can be used to organize and analyze information about the user context and, through this process, address challenges to providing usable climate information to decision-makers and other users. These challenges include moving beyond barriers to forecast use, fostering engagement with user groups, identifying use-inspired research topics, and ultimately, the development of climate products that are usable and actionable.

2.6.1 Moving beyond barriers to use

An understanding of the user context is crucial to moving beyond barriers to make use of climate information. The decision calendar approach allows RISAs to explore more sophisticated and contextual second-generation questions beyond a simple documentation of decision-makers and their stated needs, thus providing insights into cultural and institutional barriers, as well as the temporal and spatial barriers of the information itself [1–5]. Mapping out the perspectives of the broader decision context, for example, the manager's own stakeholders, can point to institutional entry points for climate information. The cases discussed demonstrate where entry points for climate information were used in endangered species recovery plans and opportunities to strategically plan the seasonal and/or geographic allocation of resources such as in the management of wildfire in the western United States. A decision calendar analysis may uncover aspects of the decision process that may not be obvious in formal operating plans, such as the need to reconcile the adaptive management process with existing annual operating plans to achieve ecological restoration [20] or the need to include other participants in the process of identifying potential uses of climate predictions [4,18].

2.6.2 Fostering engagement

A deliberate part of developing decision calendars is fostering ongoing engagement, and documenting the resulting opportunities. Ongoing engagement and partnerships in each of the cases presented have provided opportunities when key climate or societal events occurred. For example, as the drought of 2002 evolved in the western United States, understanding the decision context and calendar informed the rapid response to drought activities in partnership with water managers, which was an explicit effort in experimental services that year. Engagement included participation in water management meetings to provide climate observations and predictions, beginning early in the spring and through the water year. Interest in the climate information was heightened, including both paleoclimate and climate change data. Thus, beyond simply providing information about perceived needs, the ongoing engagement created a dialogue about climate risks and raised the level of climate literacy across time scales among decision-makers and their stakeholders.

2.6.3 Informing RISA work with stakeholders

We have used this approach to assess the needs and opportunities as the RISA program and its partners work in new regions and on new user contexts. A decision calendar approach was applied to scope uses of decadal climate information and predictability [35], and in developing projects with public lands and ecosystem managers involved in the DOI North Central Climate Science Center [Ray *et al.*, in preparation]. The decision calendar framework was recently used to identify critical information needs and lead times in a NOAA and the U.S. Army Corps of Engineers partnership to understand, explain, and assess the predictability of climate extremes in the Missouri River Basin associated with 2011 flooding. As part of an ongoing user-needs assessment in the Missouri Basin, this approach could serve as a baseline to document and understand how needs evolve over time.

2.6.4   Informing usable and actionable climate products and services

An understanding of the context described here gives insight for making a product that is useable in any of the four ways we have identified: consulted, considered, incorporated formally, or in a dialogue about risks. To be usable information, defined as science-based knowledge transformed to be readily understandable and immediately available to be incorporated into decision-making, products must respond to the temporal, spatial, and institutional needs of users. There are increasing numbers of user groups and demands for information, but by considering the decision calendars and types of climate products desired by different user groups, this analysis has provided information to climate services by identifying the common types of information needed across multiple groups.

The temporal aspect of the decision-calendar framework identifies entry points for products that were previously not available or not usable in operations, planning, and policy processes, by providing a better understanding of annual cycles of decision-making and the longer-term planning cycles they are embedded in. Analysis of the decision calendars identifies critical time periods for climate forecasts and information, as well as to infer climate information that is potentially useful, but not currently used. When desired information is not available, this analysis reveals what information is considered, what information might be important but is not considered now, and may indicate the information that would meet user needs. Thus, new products that better meet users' needs may be derived from existing products. Alternatively, if the knowledge is not available, use-inspired research may be identified. Thus, the development of decision calendars also has been an effective practice for supporting use-inspired research and guiding climate research priorities, because users needs identified can be used to drive research and product development in addition to scientists' own identification of the research questions.

Decision calendars can also help inform climate services by identifying areas where there is overlap of predictability or certainty in climate understanding with information desired by users. It may reveal new types of climate services needed, beyond the current spectrum of climate information products and services. Where products are not available, the decision calendar approach points to needs for use-inspired climate science research or applied research to develop usable science.

2.6.5   Other examples of the decision calendar approach

A decision calendar integrating adaptive management decision-making and hydroclimate information was used to identify entry points in the planning and operational decision-making processes for the Upper Colorado River and Glen Canyon Dam and to map the needs for climate-related information relative to ecologic restoration decision-making [19]. An annual agro-climate decision calendar was developed to document the influence of external forces such as the impact of ENSO on the seasonality of precipitation in Trinidad during key activity periods and to identify the types of climate information needed as well as the time frame during the decision-making process [36]. In the Central Rift Valley of Ethiopia, an agro-climate calendar was used to better understand the relationship between information needs and opportunities among farmers, resource providers, and climate forecast providers, to explore the improved use of climate information to guide agricultural decisions, and to evaluate the feasibility of tailoring seasonal rainfall forecasts for different cropping systems [37]. The annual cycle of farming and livestock production in the Arkansas Valley, Colorado, was recorded in a decision calendar, which identified the time frame for information on critical climate conditions and processes, opportunities for forecasts and other products to inform decisions and resulting benefits; simple improvements in forecasting and forecast applications were shown to be financially important [38]. In each of these studies as well as the cases presented, the development of decision calendars had value both as a collaborative process for scientists and decision-makers to engage iteratively and learn from each other in the coproduction of knowledge, as well as the decision calendar as a research product.

2.7 Conclusion

Decision calendars have proven useful in linking resources management planning processes and operational issues with potential uses of climate information and forecasts at various lead-times. Decision calendars are both a research product and an effective process for developing sustained and systematic engagement for scientists and decision-makers to interact iteratively and collaboratively. The RISAs and their partners continue to use the decision calendar approach for integrating information needs and climate science to identify entry points for climate information and forecasts, to assess the current and potential roles of climate information in policy, planning, and decision-making to manage resources and reduce the impacts, and to motivate and guide use-inspired research throughout the weather and climate science communities.

Acknowledgments

Funding to support this work came from the NOAA Climate Programs Office to the WWA and to the Climate Assessment for the Southwest, and from the NOAA Office of Oceanic and Atmospheric Research. The authors are grateful to K. Dow, H. Yocum, and two anonymous reviewers for thoughtful and constructive comments. We particularly thank the following people for discussions contributing to this work: R.S. Pulwarty (NOAA), J.L. Wescoat Jr, R. Brunner, W.T. Travis (all University of Colorado), K.T. Redmond (Western Regional Climate Center), Eric Kuhn (Colorado River Water Conservation District), Mark Waage (Denver Water), and George Smith (USFWS).

References

1 Jacobs, K.L. (2002) *Connecting Science, Policy and Decision-Making: A Handbook for Researchers and Science Agencies*, NOAA, Office of Global Programs, Silver Spring, Maryland. URL http://rts.nccmt.ca/uploads/116_304%20-%20Jacobs%20(2002).pdf.

2 Pulwarty, R.S. & Redmond, K.T. (1997) Climate and salmon restoration in the Columbia River basin: the role and useability of seasonal forecasts. *Bulletin of the American Meteorological Society*, **78**, 381–397.

3 Hartmann, H.C., Bales, R. & Sorooshian, S. (2002) Weather, climate, and hydrologic forecasting for the US Southwest: a survey. *Climate Research*, **21**, 239–258.

4 Lowrey, J., Ray, A.J. & Webb, R.S. (2009) Factors influencing the use of climate information by Colorado municipal water managers. *Climate Research*, **40**, 103–119.

5 Jacobs, K., Garfin, G. & Lenart, M. (2005) More than just talk: connecting science and decisionmaking. *Environment*, **47**, 7–21.

6 Lemos, M.C. & Rood, R.B. (2010) Climate projections and their impact on policy and practice. *WIREs Climate Change*, **1**, 670–682. doi:10.1002/wcc.71

7 Dilling, L. & Lemos, M.C. (2011) Creating usable science: opportunities and constraints for climate knowledge use and their implications for science policy. *Global Environmental Change*, **21**, 680–689.

8 Lemos, M.C. & Morehouse, B.J. (2005) The co-production of science and policy in integrated climate assessments. *Global Environmental Change*, **15**, 57–68.

9 National Research Council. (2009) *Informing Decisions in a Changing Climate*. Washington, DC: The National Academies Press.

10 McNie, E.C. (2007) Reconciling the supply of scientific information with user demands: an analysis of the problem and review of the literature. *Environmental Science & Policy*, **10**, 17–38.

11 Michaels, S. (2009) Matching knowledge brokering strategies to environmental policy problems and settings. *Environmental Science & Policy*, **12** (**7**), 994–1011. doi:10.1016/j.envsci.2009.05.002

12 Wescoat, J.L. Jr., (1991) Managing the Indus River basin in light of climate change: four conceptual approaches. *Global Environmental Change*, **1** (**5**), 381–395.

13 Lasswell, H.D. (1956) *The Decision Process: Seven Categories of Functional Analyses*. University of Maryland, College Park, MD.

14 Gunderson, L. *et al.* (1995) *Barriers and Bridges to the Renewal of Ecosystems and Institutions*. Columbia University Press, New York.

15 Cresswell, J.W. & Clark, V.L.P. (2011) *Designing and Conducting Mixed Methods Research*, 2nd edn. SAGE Publications, Thousand Oaks, CA, pp. 488.

16 Jacobs, K.L. & Pulwarty, R.S. (2003) Water resource management: science, planning and decision-making. In: Lawford, R., *et al.* (eds), *Water: Science, Policy, and Management*. American Geophysical Union, Washington, DC.

17 Ray, A.J. (2004) *Linking Climate to Multi-Purpose Reservoir Management: Adaptive Capacity and Needs for Climate Information in the Gunnison Basin, Colorado*, pp. 328. Dissertation, Department of Geography, University of Colorado.

18 Ray AJ. (2003) Reservoir management in the Interior West: the influence of climate variability and functional linkages of water. In *Climate, Water, and Transboundary Challenges in the Americas*, H Diaz, BJ Morehouse, (eds), pp. 193–217. Kluwer Press.

19 Brunner, R.D., Colburn, C.H., Cromley, C.M. & Klein, R.A. (2002) *Finding Common Ground: Governance and Natural Resources in the American West*. Yale University Press, pp. 224.

20 Pulwarty, R.S. & Melis, T. (2001) Climate extremes and adaptive management on the Colorado River. *Journal of Environmental Management*, **63**, 307–324.

21 Rogers, E.M. (1995) *Diffusion of Innovations*, 4th edn. The Free Press, New York, pp. 519.

22 Clark, M. & Hay, L. (2004) Use of medium-range weather forecasts to produce predictions of streamflow. *Journal of Hydrometeorology*, **5**, 15–32.

23 Jain, S., Woodhouse, C.A. & Hoerling, M.P. (2002) Multidecadal streamflow regimes in the interior western United States: implications for the vulnerability of water resources. *Geophysical Research Letters*, **29**, 2036–2039.

24 Garfin, G.M. & Morehouse, B.J. (2001) *Facilitating Use of Climate Information for Wildfire Decision-Making in the U.S. Southwest*. Proceedings of the AMS Fourth Symposium on Fire and Forest Meteorology, pp. 116–122. Reno, NV.

25 Corringham, T.W., Westerling, A.L. & Morehouse, B.J. (2008) Exploring use of climate information in wildland fire management: a decision calendar study. *Journal of Forestry*, **106**, 71–77.

26 Garfin, G.M. *et al.* (2003) *The 2003 National Seasonal Assessment Workshop: A Proactive Approach to Preseason Fire Danger Assessment*. Preprints, Fifth Symp. on Fire and Forest Meteorology and Second Int. Wildland Fire Ecology and Fire Management Congress, Orlando, FL, Amer. Meteor. Soc., J9.12.

27 Owen, G., McLeod, J.D., Kolden, C.D. *et al.* (2012) Wildfire management and forecasting fire potential: the roles of climate information and social networks in the Southwest United States. *Weather Climate and Society*, **4**, 90–102.

28 Higgins, W, and Gochis, D (2007) Synthesis of results from the North American Monsoon Experiment (NAME) process study. *Journal of Climate*, **20**, 1601–1607. doi: 10.1175/JCLI4081.1.

29 National Research Council. (2001) *A Climate Services Vision: First Steps Toward the Future*, National Academies Press Washington, DC.

30 Foster, J. (2008) *Crossing the Valley Of Death: The NOAA Transition Of Research Applications To Climate Services (TRACS) Program*. American Meteorological Society (AMS) 88th Annual Meeting, Extended Abstract 7A.1.URL https://ams.confex.com/ams/88Annual/techprogram/paper_132789.htm.

31 Ray, A.J., Garfin, G.M., Wilder, M. *et al.* (2007) Applications of monsoon research: opportunities to inform decisionmaking and reduce regional vulnerability. *Journal of Climate*, **20**, 1608–1627.

32 Miles, M.B. & Huberman, A.M. (1994) *Qualitative Data Analysis: An Expanded Sourcebook*, 2nd edn. SAGE Publications, Thousand Oaks, CA.

33 Bickman, L. & Rog, D.J. (2008) *The SAGE Handbook of Applied Social Research Methods*, 2nd edn. SAGE Publications.

34 Bernard, R.H. (2005) *Research Methods in Anthropology: Qualitative and Quantitative Approaches*, 4th edn. Altamira Press, New York, pp. 803.

35 Ray, A.J. (2008) Water Resources Decision-Makers and their needs for Decadal Climate Prediction, US CLIVAR Variations Newsletter, Oct 2008. http://www.usclivar.org/sites/default/files/Variations-V6N2.pdf.

36 Pulwarty, R.S., Eischeid, J. & Pulwarty, H. (2001) *The Impacts of El Niño-Southern Oscillation Events on Rainfall and Sugar Production in Trinidad: Assessment and Information Use*. Proceedings of the XXVII West Indies Sugar Technologists Conference 23–27 April, 2001 Port of Spain, Trinidad. Available on CD.

37 Walker, S., *et al.* (2002) *The Use of Agroclimatic Zones as a Basis for Tailored Seasonal Rainfall Forecasts for the Cropping Systems in the Central Rift Valley of Ethiopia*. URL http://www.climateadaptation.net/docs/papers/WalkerEthiopiaFINAL.pdf [accessed on 8 September 2015].

38 Weiner, J.D. (2004) Small agriculture needs and desires for weather and climate information in a case study in Colorado. In: *AMS 2004 Users Conference Pre-print*. American Meteorological Society, Seattle, WA.

CHAPTER 3

Climate science for decision-making in the New York metropolitan region

Radley Horton[1,2], Cynthia Rosenzweig[1,2], William Solecki[3], Daniel Bader[1,2] and Linda Sohl[1,2]

[1] Center for Climate Systems Research, Columbia University Earth Institute, 2880 Broadway, New York, NY 10025, USA
[2] NASA Goddard Institute for Space Studies, 2880 Broadway, New York, NY 10025, USA
[3] Department of Geography, CUNY Institute for Sustainable Cities, Hunter College, 695 Park Avenue, New York, NY 10065, USA

3.1 Introduction

New York City is one of the world's most vulnerable cities to coastal flooding, due to a high concentration of population and assets near a coastline exposed to warm-season tropical storms and cold-season Nor'easter storms [1]. Among U.S. cities, New York City is second only to New Orleans in population living less than 4 ft above the local high tide [2]. By the 2050s, average annual losses due to coastal flooding alone could exceed $2 billion for the combined New York City—Newark region [1]. Perhaps the most iconic example of a vulnerable New York City asset is the financial district located at the southern tip of Manhattan, however low-lying coastal assets include the full complement of major highways, subways and tunnels, hospitals, schools, wastewater treatment plants, food distribution centers, and people's homes [3]. Given the magnitude of the assets at risk, a compelling case can be made that long-term adaptation makes economic sense for New York City. Given New York's access to economic, human, and technological resources for resilience measures, the City may be able to achieve this resilience. The city's political environment—New York City is a place where climate science is generally not a partisan issue—and the city's experience with uncertainty and *overall* risk framing (e.g., financing of bond issues for multi-billion dollar infrastructure with multidecade expected lifetimes), encourage *climate* risk framing.

Climate in Context: Science and Society Partnering for Adaptation, First Edition.
Edited by Adam S. Parris, Gregg M. Garfin, Kirstin Dow, Ryan Meyer, and Sarah L. Close.
© 2016 John Wiley & Sons, Ltd. Published 2016 by John Wiley & Sons, Ltd.

While New York City's size and economic resources currently make it nearly unique in some respects, it also epitomizes many of the hazards faced by other cities in the Northeast, the United States, and the world. Like many cities, New York City owes its location to the historical need to be situated near rivers and the coast. As the world's population becomes increasingly urban and increasingly coastal, more and more cities are rivaling and surpassing New York in size and exposure to climate risk.

Cities big and small, coastal and inland, face multiple climate and nonclimate stressors. For example, climate change hazards will interact with other urban hazards such as the urban heat island and poor air quality [4]. This chapter relates how climate scientists and decision-makers in New York City have worked together to make New York City more resilient in the face of these climate hazards. It is organized chronologically, covering: (1) a background on climate assessment in the region, (2) science–policy interactions after Mayor Bloomberg convened a climate change task force in 2008, (3) outcomes of those interactions, (4) how Hurricane Sandy changed the playing field in late 2012, (5) the critical role played by the Consortium for Climate Risk in the Urban Northeast (CCRUN) National Oceanic and Atmospheric Administration (NOAA)-Regional Integrated Sciences and Assessments (RISA) during Sandy and in post-Sandy initiatives, (6) outcomes of those initiatives, and (7) future needs.

3.2 Background (late 1990s to 2007)

There is a long history of climate assessment in the New York City metropolitan region (Table 3.1). Engagement with decision-makers was a core tenet of these assessments from the very beginning. Representatives from relevant agencies and groups were participating actively in the Metropolitan East Coast (MEC) Assessment sector teams from the outset late in the last century. The Environmental Protection Agency Region 2, Port Authority of NY/NJ and FEMA Region 2 were major stakeholders for the 2001 Report "Climate Change and a Global City" [5], providing critical framing regarding "challenges and opportunities" throughout the assessment. Early leadership was also shown by New York City's Department of Environmental Protection (NYCDEP), which in 2003 began a major study of how climate change could influence water supply and water treatment [6], culminating in the Agency's Climate Action Plan [7]. The Metropolitan Transportation Authority developed a Blue Ribbon Commission on Sustainability that considered greenhouse gas mitigation options and the need to reduce vulnerabilities to climate change [8]. These initiatives were grounded in climate science

Table 3.1 Climate assessment reports for the New York metropolitan region.

Year	Report title	Organization/Publisher
2013	New York City Panel on Climate Change Climate Risk Information 2013	Columbia University and CUNY
2011	New York State ClimAID Adaptation Assessment	New York State Energy Research & Development Authority
2010	New York City Panel on Climate Change	Columbia University and CUNY
2009/2010	Long Island Shore Study	The Nature Conservancy
2008	New York City's Vulnerability to Coastal Flooding: Storm Surge Modeling of Past Cyclones	Bulletin of the American Meteorological Society
2008	Climate Change Program Assessment and Action Plan	New York City Department of Environmental Protection
2007	Confronting Climate Change in the U.S. Northeast: Science, Impacts and Solutions	Union of Concerned Scientists
2007	August 8, 2007 Storm Report	Metropolitan Transportation Authority
2001	Climate Change and a Global City: An Assessment of the Metropolitan East Coast Region (MEC)	U.S. National Assessment & Columbia Earth Institute
1999	Hot Nights in the City: Global Warming, Sea Level Rise and the New York Metropolitan Region	Environmental Defense Fund
1996	The Baked Apple? Metropolitan New York in the Greenhouse	New York Academy of Sciences

and impact and adaptation assessment provided by Columbia University and other area universities. The involvement of some CCRUN researchers spans this entire history and predates the formation of CCRUN as a RISA.

In 2007, Mayor Michael Bloomberg launched an ambitious city-wide sustainability initiative known as PlaNYC 2030. While principally focused on greenhouse gas mitigation and sustainability issues, climate risk and adaptation were also briefly addressed in the initial PlaNYC Report [9]. In 2008, New York City ramped up its climate risk and adaptation assessment efforts, forming the Climate Change Adaptation Task Force (CCATF) comprising approximately 40 city agencies and private-sector companies responsible for critical infrastructure systems region-wide. The Mayor also convened the first New York City Panel on Climate Change (hereafter "NPCC1"), a scientific advisory group comprising climate scientists as well as risk management and insurance experts. The NPCC1 was charged with providing information about climate hazards, impacts, and adaptation to inform New York City decision-making. Key elements of this region-wide effort led by New York City's government include:

1 *Strong leadership*—Mayor Bloomberg attended the first meeting of the Task Force, signaling to each CCATF member agency the importance of climate risk management and adaptation to his administration.
2 *Oversight by a coordinating body*—The Mayor's Office of Long Term Planning and Sustainability (OLTPS) answered directly to the Mayor, rather than being housed within an existing agency; this reduced the risk of competition between agencies, and elevated the initiative.
3 *Distinct NPCC and CCATF processes*—Frequent interactions between the two bodies helped ensure the integrity of the science-stakeholder process.

With this structure in place, the NPCC1 and the CCATF were well-positioned to cogenerate climate risk information (CRI) for New York City.

3.3 Science policy interactions during NPCC1 (2008–2009)

Under Mayor Bloomberg's leadership and the oversight of the OLTPS, direct interactions between the NPCC1 and the CCATF were encouraged. The CCATF included four infrastructure sector workgroups: Communications, Energy, Transportation, and Water & Waste, as well as an overarching Policy group. As climate projections were being developed, NPCC1 scientists met multiple times with each sector workgroup, asking questions aimed at understanding the decision context such as "What weather and climate events most impact your systems today?" and "What are the potential weak points in your systems that concern you the most?" [10,11]. Interactions between the NPCC1 and the CCATF extended beyond the topic of climate hazards to include: (1) filling out risk matrices based on the combination of likelihood of occurrence of a climate-related impact (e.g., flooding of a tunnel) and magnitude of consequences should that impact event occur, and (2) brainstorming possible adaptation strategies and identifying evaluation criteria to inform selection of specific strategies. These adaptation evaluation criteria included feasibility of implementation, effectiveness if implemented, peripheral cobenefits and costs, and resiliency (defined as flexibility or changeability of the adaptation as new information becomes available) [11].

The results of these discussions directly informed the climate science products developed. For example, stakeholder feedback about their decision context influenced several elements of the CRI generated by the NPCC1:

1 *Identification of major climate hazards.* With its continental climate and coastal location, New York City is exposed to a broad set of climate hazards (Box 3.1). These include high and low extremes of temperature, drought and intense precipitation, coastal flooding, wind events, and winter weather. Inclusion of a diverse set of infrastructure managers, some with a long history of service, helped insure that a broad set of climate hazards were analyzed, including some, such as drought, that have not been experienced to a serious extent since the 1960s. Decision-makers highlighted the importance of focusing on climate extremes in addition to changes in annual mean temperature, precipitation, and sea level, since the greatest system impacts tend to be associated with extremes.

Box 3.1 Climate hazards in the New York City region.

Heat waves. It is becoming increasingly clear that high temperature extremes are a major public health hazard, representing the biggest weather-related cause of deaths [12] in the U.S. In New York City, a strong urban heat island effect (measured at ~4.5 °F), potential for high humidity during heat waves, and poor air quality can all combine to make heat waves particularly dangerous [13,14].

Cold air outbreaks. The winter of 2013–2014 served as a reminder that extreme cold can occur in New York City as well. Preliminary analysis suggests that during the cold season of 2013–2014, New York City (Central Park) experienced 92 days with temperatures at or below freezing, the most since 1976–1977. Cold air outbreaks increase energy expenditures on heating, damage infrastructure such as aging water mains, and are associated with elevated mortality [15]. While the projected reduction in cold air events may make for less harsh conditions in the winter, cold air outbreaks can be critical for urban ecosystems, for example, by killing off insect pests that cause a nuisance, or major health and economic impacts, later' in the year.

Intense rainfall events. Heavy precipitation events, which have become more common in New York City and the Northeast as a whole [16,17], have a variety of adverse impacts on New York City. An August 2007 "surprise" rain event during the morning rush hour dropped between 1.5 and 3.5 inches in approximately 2 hours, disrupting 19 subway line segments [18]. Heavy rain events also overwhelm the city's combined sewer overflow system, leading to discharge of untreated sewage and other pollutants into the region's waterways, and increase turbidity due to excess sediment in drinking water reservoirs [7].

Coastal storms. New York City is highly vulnerable to coastal storms, which cause coastal storm surge, inland flooding, and wind damage. While the greatest coastal flood heights at multidecadal timescales are associated with warm-season tropical cyclones, at shorter timescales coastal flooding is dominated by cold-season Nor'easters. Nor'easter effects can be large, in part because they can be slow-duration events that persist for multiple tidal cycles, thereby compounding flooding effects associated with both precipitation and storm surge. Impacts of coastal storms on the built environment can be profound and extend beyond the built environment to include salt-water intrusion into groundwater and mobilization of polluted sediments [10].

High winds. Wind events can be caused by thunderstorms and coastal storms. Dangers include broken windows, especially atop tall buildings where wind speeds are higher during tropical storms, and collapse of unsecured equipment and tree limbs. Both of these climate hazards have led to fatalities in New York City in recent years.

Snow and ice. Snow and ice storms are capable of paralyzing the New York City region. In addition to transportation sector hazards, other risks include roof collapses, fires, and other damage associated with road salt percolation into electrical systems below the ground. Road treatment is also a significant expense for New York City. In the borough of Manhattan, there are 1071 pieces of snow removal equipment, 6 salt-storage facilities, and 10 storage tanks for calcium chloride [19]. In some parts of New York City's outer boroughs, power systems are above ground, increasing the risk of failures associated with ice storms and heavy snow.

Drought. Drought is generally considered a relatively minor risk for New York City, given the region's abundant rainfall during a normal year, large reservoir storage capacity, and reductions in water usage in recent decades. Nevertheless, the drought of record of the 1960s shows that there is a precedent for water supply limitations, and therefore a need for adaptation measures to be ready for drought impacts, should they occur.

2 *Climate variable thresholds*: For those climate variables where the NPCC1 found there was sufficient information available to justify quantitative projections (such as high/low temperature, intense precipitation, and coastal flood recurrence), stakeholders selected the specific thresholds to be used in consultation with the NPCC. These thresholds represented key factors in the local context. For example, days with maximum temperatures at or above 90 °F were selected since New York City residents associate 90 °F with high heat, and it is the threshold used by the National Weather Service for heat wave definition in New York City. The 1-in-100 year coastal flood was emphasized because it is a foundation for regulations including building codes, zoning, and insurance.

3 *Model-based distribution*: As climate science and adaptation assessment have advanced, it has become increasingly common for decision-makers to consider multiple climate and greenhouse gas emissions scenarios. The NPCC1 conducted analyses with the 16 Coupled Model Intercomparison Project 3 [20] Global Climate Models (GCMs) developed in support of the 2007 Intergovernmental Panel on Climate Change (IPCC) Report that provided projections for three different greenhouse gas (GHG) emissions scenarios [21]. It was emphasized to decision-makers that using multiple GHG scenarios samples a range of possible future concentrations of greenhouse gases, the key drivers of climate change over current planning horizons, and that using multiple GCMs samples the range of sensitivity of the climate system to a given greenhouse gas concentration. Projections were initially presented as histograms representing the full distribution; however, for ease of understanding, stakeholders decided to instead emphasize the projections associated with two points in the distribution: the 17th and 83rd percentiles (the central range). Box 3.2 describes a case where stakeholder risk considerations led to consideration of more extreme points in the distribution.

Box 3.2 The high-end "rapid ice melt scenario" and risk-averse infrastructure stewards.

Managers of long-term mission-critical infrastructure, such as the Port Authority of New York and New Jersey, which manages bridges, tunnels, and airports, expressed a need for low probability, high impact scenarios in addition to those bounded by the central range. In response, the NPCC developed a rapid ice-melt scenario that considered the possibility of extensive dynamical melting of land-based ice [22,23]. The Port Authority requested additional meetings with NPCC1, including a specialized meeting with their Aviation Division, to discuss these high-end scenarios and how they could be incorporated into planning for the protection of long-lived critical infrastructure, such as the region's three low-lying airports.

4 *Timeslices*: The time periods of analysis were selected based on decision-maker feedback on key issues in their context. Specifically, groups who replaced infrastructure relatively frequently, such as telecommunications providers, focused on the near-term period, defined as the 2020s. Those charged with managing long-lived infrastructure such as water tunnels (NYCDEP), bridges and transportation tunnels (Port Authority of New York and New Jersey and the Metropolitan Transportation Authority) were also interested in the long term, defined as the 2080s. While stakeholders managing short-lived infrastructure requested even nearer-term projections, the researchers responded that projections for a period earlier than the 2020s are not recommended, since 30-year timeslices (e.g., defining the 2020s as 2010–2039) are required to discern the predictable climate change signal from the unpredictable multiannual "noise" associated with natural climate variability. At point locations, decadal climate variability can be significant for some variables, such as precipitation. This example illustrates that, on some occasions, climate scientists need to resist the inclination to meet decision-maker requests that exceed the state of our scientific understanding.

5 *Qualitative information*: Projected directional changes in extreme events with associated likelihoods were developed in response to scientifically challenging stakeholder requests. The hazards included in these analyses were heat index, ice storms/freezing rain, snowfall, downpours, lightning, intense hurricanes, Nor'easters, and extreme winds. Even though it was not possible, given the state of the science to make quantitative projections for these extreme events, their impacts on infrastructure and provision of services was sufficient to justify their inclusion as qualitative CRI.

6 *Presentation of information*: The NPCC1 worked with the OLTPS and CCATF to present climate information in the manner most useful to a range of decision-makers. This led to two primary products: (1) a short two-page "tear-sheet" that contains the core climate projections, and (2) a lengthy report that provides the risk management and climate science context, the details behind the methods used to make the projections, caveats and key uncertainties, and paths forward to improved projections in the future [24]. The "tear sheet" was intentionally designed to concisely present a common denominator of standardized projections that could be used in a coordinated way by different city agencies in their planning. The lengthy report provides details that might be of interest to those wanting to better understand the science, explore more sector-specific applications, or gain in-depth perspectives. As examples of the latter, the report noted that New York City's temperatures by the 2050s might be similar to Norfolk, Virginia today, and that the warmest years experienced today are projected to be among the coldest years experienced in the 2050s.

3.4 The post-NPCC1 landscape (2009–2012)

As a result of the combined CCATF and NPCC1 initiatives, New York City decision-makers embarked on several adaptation initiatives between 2009 and 2012. These included: (1) planting over 300 "Greenstreets" with vegetation that absorbs stormwater, (2) securing citywide high-resolution Light Detection and Ranging (LiDAR) elevation data, which helps to identify the areas most vulnerable to coastal flooding, (3) incorporating sea level rise into the City's Comprehensive Waterfront Plan, and (4) launching enhanced emergency response and preparedness programs [3,25].

Another important step in mainstreaming climate information into New York City decision-making was the September 2012 codification of the NPCC into the City's local laws [26]. This requirement that the NPCC be reconvened periodically to update the climate projections helped institutionalize the science, so that future assessments are more likely to be conducted in a manner that builds on prior work.

The CCRUN RISA was launched in October 2010, during the period between NPCC1 and Hurricane Sandy (Box 3.3). CCRUN's Principal Investigator led the first NPCC assessment, and CCRUN was formulated to address a broader set of issues—over the larger Philadelphia to Boston corridor—than were addressed in NPCC1 for New York City. By including five universities, CCRUN was able to conjoin additional sector expertise, including public health, hydrodynamic "storm surge" modeling, green infrastructure, and social vulnerability. As Hurricane Sandy would sadly reveal, all these skill sets, applied over a larger spatial area, would be essential.

Box 3.3 The Consortium for Climate Risk in the Urban Northeast (CCRUN)

The CCRUN conducts stakeholder-driven research that reduces climate-related vulnerability and advances opportunities for adaptation in the urban Northeast. CCRUN is designed to address the complex challenges that are associated with densely populated, highly interconnected urban areas, such as urban heat island effects; poor air quality; intense coastal development; multifunctional settlement along inland waterways; complex overlapping institutional jurisdictions; integrated infrastructure systems; and highly diverse, and in some cases, fragile socio-economic communities.

CCRUN's projects are focused in four broad sectors: Water, Coasts, Health, and Green Infrastructure. Research in each of these sectors is linked through the cross-cutting themes of climate change and community vulnerability. CCRUN's stakeholder-driven approach to research supports investigations of the impacts of a changing climate, population growth, and urban and economic policies on the social, racial, and ethnic dimensions of livelihoods and of communities in the urban Northeast. Disadvantaged socio-economic groups have been particularly underserved in the area of climate change, and one of CCRUN's long-term goals is the building of adaptive capacity among such groups to current and future climate extremes.

3.5 Hurricane Sandy (29 October 2012)

Hurricane Sandy was one of the largest tropical storms ever to strike the New York metropolitan region, with a radius of tropical storm winds extending out almost 500 miles (Figure 3.1). Due to the large size of the storm and the fact that its peak storm surge coincided with high astronomical tides, the high water levels associated with Sandy far surpassed the levels during prior storms in New York City and New Jersey (see Table 3.2). This extreme surge led to severe impacts for the region, even though Sandy's central winds and rainfall were lower than is often the case during tropical cyclones.

New York City.

Figure 3.1 Satellite image of Hurricane Sandy on 29 October 2012. Source: NASA.

These impacts include more than 40 lives lost in the city, approximately $20 billion in damages, electrical outages for as many as four million in the region, and major flooding of three vehicular tunnels and seven rail lines under the East River.

In the aftermath of Hurricane Sandy, it became clear that the storm's large impacts on the city would elevate the role of climate science in

Table 3.2 Top 10 coastal flood heights at The Battery, New York, NY (past 74 years).

Storm	Date	Water level (Mean Lower Low Water (MLLW)), ft	Water level (North Atlantic Vertical Data (NAVD)), ft
Hurricane Sandy	29 October 2012	13.88	11.1
Hurricane Donna	12 September 1960	10	7.22
Nor'easter	11 December 1992	9.7	6.92
Hurricane Irene	28 August 2011	9.51	6.73
Nor'easter	25 November 1950	9.12	6.34
Ash Wednesday Storm	6–7 March 1962	8.92	6.14
Nor'easter	2 January 1987	8.84	6.06
Halloween ("Perfect Storm")	31 October 1991	8.73	5.95
Blizzard of '84	29 March 1984	8.53	5.75
Nor'easter	2 January 1987	8.38	5.6

New York City's decision-making. While prior assessments had pointed to many of the city's vulnerabilities (e.g., the likelihood of flooding of key transportation infrastructure) [27], seeing these events come to pass at such magnitude led to a call for action by Mayor Bloomberg. Furthermore, while many impacts had been predicted by prior assessments, Sandy highlighted additional, unforeseen vulnerabilities. For example, dependencies across infrastructure systems led to shortages of gasoline, and social services for vulnerable populations were severely hindered by flooding of key facilities such as hospitals and elder-care facilities.

Several aspects of CCRUN's organizational design, experience, and history made it well positioned to address pressing needs that became evident post-Sandy. First, Sandy revealed a need for science in place, and science in time [28]. Because CCRUN's researchers were already focused on Sandy-relevant issues (e.g., coastal modeling and community vulnerability; [29]) and relationships were already in place with decision-makers, CCRUN had a process in place to respond to local needs immediately. Second, Sandy revealed that interdisciplinary teams were essential to tackle the complex problems that arose. For example, Sandy showed how a combination of sea level rise, storm surge, public health considerations, and social vulnerability led to devastating community impacts. CCRUN was ready to engage immediately since it included existing partnerships of climate, hydrodynamic, public health, and social science researchers. Third, Sandy revealed the importance of "having the climate conversation" about risk, vulnerability, and solutions—both mitigation and adaptation—during a teachable moment. CCRUN researchers had extensive experience leading such dialogues in the region. Finally, Sandy revealed the importance of considering impacts across multiple jurisdictions and regions, which CCRUN was well positioned to do given its broader focus on the Boston to Philadelphia urban corridor.

3.6   CCRUN's role in Sandy response

In the days leading up to and during Sandy, CCRUN researchers communicated risks through prominent national media and offered storm-surge forecasts. CCRUN researchers also tracked and modeled critical hyper-local meteorological data during the storm, including (1) wind gusts and (2) water elevation (Figure 3.2). CCRUN researchers also played key roles in longer-term response, which is the focus of this section.

Coincidentally, CCRUN's annual meeting was held the week after Hurricane Sandy, when some team members were still without power. During that meeting—as both the extent of the societal needs and the number of complex, unanswered questions became clear—CCRUN decided to turn much of its focus for the next year to a Sandy Research Agenda. Key elements included the following, organized by sectoral team:

- *Coastal zones:* Develop new metrics of coastal storm risk that better communicate risk, for example, how can we translate "storm-surge height" into "flood danger" or other human-scale metrics; including danger from waves as well as surge.
- *Weather and climate:* Research the influence of large-scale changes on hurricane strength (e.g., changes in North Atlantic Ocean temperatures) and path (e.g., changes in large-scale atmospheric steering patterns; [30]).
- *Water resources:* Improve and update the intensity–duration–frequency (IDF) curves used to design storm water and sewer systems to better project the occurrence of extreme precipitation events associated with hurricanes and other storms.
- *Public health:* Explore new measures to reduce both the immediate and longer-term health impacts of storm events, such as mold-related respiratory disease, gastrointestinal infections from contaminated water, and a range of mental health outcomes.
- *Vulnerability and adaptation:* Study the determinants of social resilience to coastal flooding in urban neighborhoods to build an understanding of why vulnerability and resilience differ across communities that experience the same flood levels.

CCRUN initiatives in support of NPCC, which is the subject of the next section, were also critical.

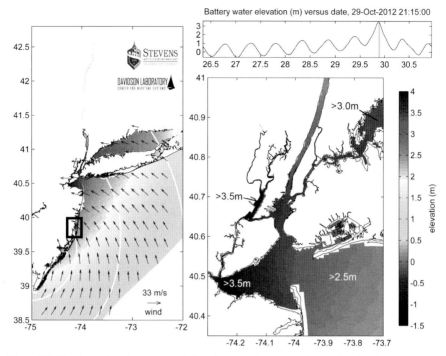

Figure 3.2 Hurricane Sandy surge modeling. Source: Stevens Institute of Technology.

3.6.1   NPCC2 and SIRR

After Hurricane Sandy, Mayor Bloomberg reconvened the NPCC ("NPCC2," hereafter) in January 2013 to provide updated CRI in support of the Special Initiative for Rebuilding and Resiliency (SIRR). SIRR was established with the objectives of: (1) analyzing the impacts of Hurricane Sandy on buildings, infrastructure, and the public; (2) assessing the risks that climate change presents to New York City in the medium-term and long-term range; (3) creating strategies to increase climate resilience throughout the city; and (4) developing proposals for rebuilding the areas of the city most damaged by Hurricane Sandy.

The SIRR was an ambitious endeavor; it entailed a significant expansion of climate adaptation efforts by the City on a short timeline. The work of the first New York City CCATF of 2008–2010 focused primarily on infrastructure, while this time the reach was broadened to speak more fully to other central elements of PlaNYC [31] including health, vulnerable populations, and natural systems. Another difference was that the city would produce a large report that documented the impacts of Sandy and included recommendations for many adaptation strategies.

An additional challenge was that the initiative would be a rapid assessment, in order to ensure that science would be provided in sufficient time to inform decisions about how to rebuild, and anticipate the arrival of funds for rebuilding. The NPCC2 was charged to create a 2013 NPCC Climate Risk Information Report with new climate projections based on the latest GCMs, observations, and new physical understanding, as well as updated coastal flood risk maps for New York City. The work of the NPCC2 was vetted with a core group of agency stakeholders, via the OLTPS.

Along with the SIRR report, the NPCC2 Climate Risk Information Report was released in June 2013[3]. The NPCC2 CRI Report builds upon the foundation established by NPCC1, although it includes some important updates and modifications. NPCC2 uses the latest GCM outputs (CMIP5) to present projections of changes in mean temperature and precipitation (Table 3.3), as well as changes in extreme events (Table 3.4). The NPCC2 developed a state-of-the-art six-component probabilistic approach to local sea level rise projections that blends model-based and observational approaches (Figure 3.3) in a significant advancement over the original NPCC approach to sea level rise. In addition, for the first time NPCC2 included

Table 3.3 Mean annual changes.

	Low estimate (10th percentile)	Middle range (25th–75th percentile)	High estimate (90th percentile)
(a) Temperature [baseline (1971–2000) 54 °F]			
2020s	+1.5 °F	+2.0 to +2.9 °F	+ 3.2 °F
2050s	+3.1 °F	+4.1 to +5.7 °F	+6.6 °F
2080s	+3.8 °F	+5.3 to +8.8 °F	+10.3 °F
(b) Precipitation [baseline (1971–2000) 50.1 in.]			
2020s	−1%	+1 to +8%	+10%
2050s	+1%	+4 to +11%	+13%
2080s	+2%	+5 to +13%	+19%
(c) Sea level rise [baseline (2000–2004) 0 in.]			
2020s	2 in.	4–8 in.	10 in.
2050s	8 in.	11–21 in.	30 in.
2080s	13 in.	18–39 in.	58 in.

Panels (a) and (b): Based on 35 GCMs and two representative concentration pathways. Baseline data are for the 1971–2000 base period and are from the NOAA National Climatic Data Center (NCDC). Shown are the low-estimate (10th percentile), middle range (25th–75th percentile), and high-estimate (90th percentile). Panel (c): Projections are based on a six-component approach that incorporates both local and global factors. The model-based components are from 24 GCMs and two representative concentration pathways. Shown are the low-estimate (10th percentile), middle range (25th–75th percentile), and high-estimate (90th percentile). Projections are relative to the 2000–2004 base period.

Table 3.4 Future Coastal Flood Events Associated with Projected Sea Level Rise.

	Low estimate (10th percentile)	Middle Range (25th–75th percentile)	High estimate (90th percentile)
2080s			
Annual chance of today's 100-year flood (1%)	2.0%	2.4–7.1%	18.5%
Flood heights associated with 100-year flood (stillwater + wave heights) (15.0 ft)	16.1 ft	16.5–18.3 ft	19.9 ft

Shown are the low-estimate (10th percentile), middle range (25th–75th percentile), and high-estimate (90th percentile) 10-year mean values from model-based outcomes.
Flood heights are derived by adding the sea level rise projections for the corresponding percentiles to the baseline values. Baseline flood heights associated with the 100-year flood are based on the stillwater plus wave heights. Flood heights are referenced to the NAVD88 datum.

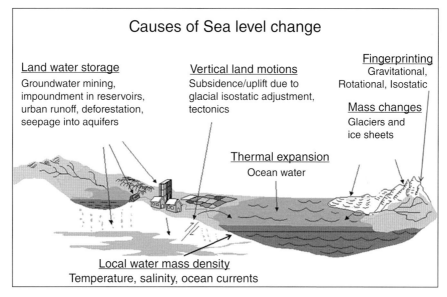

Figure 3.3 Components of sea level rise. Source: Center for Climate Systems Research, Columbia University.

quantitative information about how humidity is projected to change, based on the growing understanding among the scientists and health department stakeholders of the large health impacts of heat stress. Drought received less emphasis, due to stakeholder and scientist feedback. Specifically, it has become clearer that only multiyear droughts pose a threat to New York City's water supply, and that water usage trends and climate model projections of increasing cold-season precipitation may reduce the risk in the future. While multiyear drought risk remains a threat due to (currently) unpredictable natural variability, it is generally seen as less of a threat to New York City's water supply system than the water quality and flooding issues associated with too much rainfall.

Other changes relative to NPCC1 include use of a more extreme high-end climate forcing scenario (RCP8.5 instead of SRES A2). While the lower bound forcing scenario is similar to that used in NPCC1 (RCP4.5 instead of SRES B1), the range or results expands toward the end of the century due to a combination of the more extreme high-end scenario and availability of more GCMs (35 instead of 16). One subtle change was a stakeholder-requested shift in the central range percentiles emphasized in the report: while NPCC1 highlighted the 17th and 83rd percentiles (and showed the minimum and maximum), NPCC2 highlighted the 25th and 75th (and showed the 10th and 90th). These stakeholder-driven changes served to slightly narrow the range of results shown relative to the range emphasized in NPCC1. The NPCC2 considered updating the climate baseline from 1971–2000 to 1981–2010 but based on New York City stakeholder feedback ultimately decided not to, in order to preserve greater continuity with the NPCC1 projections stakeholders had grown accustomed to. Continuity was also maintained in presentation; NPCC2, like NPCC1, contained need-to-know information in "tear sheets," and offered broader context in an accompanying report.

For decision-makers, perhaps the most resonant NPCC2 finding is how seemingly small shifts in average conditions can profoundly shift the frequency and intensity of extreme events. These results, including the projected large increase in coastal flood frequency by the end of the century even if storms themselves do not become any stronger, inform long-term decision-making post-Sandy as well as additional NPCC2 activities. Since the release of the 2013 Report, the NPCC2 has advanced additional research related to hydrodynamic modeling, mapping, human health, and indicators and monitoring. CCRUN members have led many of these efforts, which will be published as peer-reviewed chapters in a New York Academy of Sciences book volume [32]. The NPCC2 has tapped the national RISA network (i.e., other RISA teams) in order to have the material peer-reviewed by leaders at the climate science/decision-making interface.

Other CCRUN activities of relevance to Sandy response in the region include coordinating with NOAA and other agencies such as the Army Corps of Engineers and FEMA in integrating NPCC2 projections into NOAA's Sea Level Rise Viewer, the Army Corps of Engineers' Sea Level Change Calculator, and the Sea Level Rise Tool for Sandy Recovery [33].

This initiative has the potential to encourage agencies such as FEMA to more fully integrate climate risk management into their own decision-making, while emphasizing the need for a coordinated approach to climate scenarios at local, state, and federal levels. Like other RISAs, CCRUN also contributed to the NOAA-led Sea Level Rise Technical Input Report [34] and other National Climate Assessment chapters and products.

3.7   Outcomes of NPCC2 and SIRR

The scope of the SIRR is more ambitious than prior efforts to develop resilience in New York City: Over 250 adaptation initiatives have been proposed, many of which also support greenhouse gas mitigation and sustainability objectives. It is estimated that $20 billion is needed for the entire SIRR plan. Implementing these efforts also requires ambitious engagement efforts; already 35 government agencies, 65 public officials, 320 community organizations, and thousands of individuals have been involved in the process.

As an example of the high level of detail in the SIRR, Box 3.4 shows proposed activities in a single sector, buildings.

Box 3.4 Building sector adaptation initiatives from the SIRR.

Strategy: Strengthen new and substantially rebuilt structures to meet the highest resiliency standards moving forward through:
- Improving flood resiliency regulations
- Rebuilding/repair from Sandy damage
- Zoning changes
- Housing design competition
- Buyout program with NYS
- Wind resiliency measures in building code (new buildings)

Strategy: Retrofit as many buildings as possible so that they will be significantly more resilient than they are today through:
- Incentivizing adoption of flood resiliency measures
- Establishing Community Design Centers for assisting property owners
- Retrofitting public housing to increase resiliency
- A sales tax abatement program
- A technology design competition
- Clarifying regulations for landmarked structures
- Integrating wind resiliency measures into the building code (existing buildings)
- Amending Construction Codes and best practices to prevent utility service disruptions

Table adapted from New York City Special Initiative for Rebuilding and Resiliency [3].

In 2013 alone, New York City passed 16 laws related to floodplain building codes, which offered protection against future risks associated with wind, water, and associated power failures. Most notably, FEMA's updated flood insurance rate maps were adopted with additional "freeboard" safety standards for some residences, based on NPCC2 sea level rise projections [33].

Post-Sandy recovery efforts and the SIRR have also catalyzed a range of other initiatives in the region. For example, Consolidated Edison, the primary electric, natural gas, and steam utility for the New York metropolitan region, agreed to take steps to incorporate information about extreme temperature, humidity, flooding, and winds into their planning, in order to make sure that their infrastructure could withstand projected changes in these variables. Storm hardening, demand response, and distributed generation are among the strategies to be used. Consolidated Edison agreed to undertake these measures as part of its rate filing for 2014, and this agreement was incorporated into an order issued by the company's principal regulator, the New York Public Service Commission [35]. Other initiatives include the Jamaica Bay Science and Resilience Institute funded by the City of New York, Department of the Interior and National Park Service, and the Rebuild by Design Competition organized by the Hurricane Sandy Rebuilding Taskforce and supported by the Department of Housing and Urban Development.

3.8   Future needs

While New York City has made great strides in reducing its vulnerability to climate variability and change, Sandy has revealed that much remains to be done. Furthermore, even as the City responds to the last extreme event, it will be critical that recommendations from the SIRR and CCRUN to address other types of climate hazards, such as extreme heat events and intense precipitation, are advanced as well. These recommendations include consideration of low probability, high-risk abrupt climate changes that are difficult to project using climate models, such as potential regional impacts of rapid loss of Arctic sea ice [36].

It is also critical that New York City continues to show leadership in greenhouse gas mitigation efforts. New York City's carbon footprint declined by 12.9% from 2005 to 2009, and the city appears to be on a trajectory that would enable it to meet its intermediate term targets of 30% emissions reductions from 2005 levels, by 2030. However, as the work of the NPCC reveals, deeper cuts are needed by the 2050s and beyond in order to avert the worst impacts of climate change [37].

Because New York City has developed climate science expertise within its agencies, a bottom-up approach is contributing to climate action as well. For example, New York City DEP has co-located an in-house hydrological modeling team with climate researchers [38]. This mainstreaming of climate science information within agencies and cities is a sign that climate adaptation assessments have successfully built capacity (see Chapter 11). A key future question is how to ensure that agencies continue to engage with climate scientists, and that agencies remain up to date on the latest advances in climate science understanding and climate model outputs.

Sandy's impacts have also highlighted the need for coordination across different scales of government, ranging from local communities to states and nations. Challenges include how to integrate different climate projections and how to coordinate adaptation strategies to ensure that they do not operate at cross-purposes. Cities like New York have, in many ways, led the way in implementing adaptation, and New York City is becoming increasingly vocal in national discussions about climate risk and infrastructure investment. Nevertheless, the scope and scale of the challenge point to the need for federal leadership and funding for adaptation.

As leaders at the science–policy interface, New York City decision-makers and CCRUN scientists have both the opportunity and the responsibility to share scientific products and lessons learned with other cities that may not be as far along in the process and/or have fewer resources available. A key research question is what aspects of the NYC approach (e.g., adaptation strategies) can and cannot be applied to other locations. For example, recommending capital-intensive coastal policy options that might be feasible for parts of New York City—such as large-scale coastal engineering fortifications—might be irresponsible for at-risk low-lying cities that lack economic resources or fail-safe coastal evacuation routes; for such cities especially, coastal retreat is likely to become a large part of the discussion in the coming decades.

Acknowledgments

We acknowledge the researcher team and stakeholders within the Consortium for Climate Risk in the Urban Northeast network. We also acknowledge and thank NOAA's RISA Program, and the City of New York for providing funding and other support for the initiatives described here.

Glossary

CCATF	Climate Change Adaptation Task Force
CCRUN	NOAA RISA Consortium for Climate Risk in the Urban Northeast
CMIP5	Coupled Model Intercomparison Project
CRI	Climate Risk Information
DSNY	The City of New York Department of Sanitation
FEMA	Federal Emergency Management Agency
GCMs	Global Climate Models
GHG	Greenhouse Gas
IPCC	Intergovermental Panel on Climate Change
LIDAR	Light Detection and Ranging
MEC	Metro East Coast Report, in support of First National Climate Assessment
MTA	Metropolitan Transportation Authority
NOAA	National Oceanic and Atmospheric Administration
NPCC	New York City Panel on Climate Change
NYCDEP	New York City Department of Environmental Protection
OLTPS	New York City Mayor's Office of Long-Term Planning and Sustainability
PlaNYC	New York City's Sustainability and Resiliency Plan
RCP	Representative Concentration Pathway
RISA	Regional Integrated Sciences and Assessments
SIRR	Special Initiative for Rebuilding and Resiliency

Disclaimer

This chapter reflects work and findings through approximately the middle of 2014. Given rapid advances in resilience and scientific understanding in the region, this chapter may not reflect all the latest initiatives and science developments, including updates to the NPCC projections.

References

1 Hallegatte, S., Green, C., Nicholls, R.J. & Corfee-Morlot, J. (2013) Future flood losses in major coastal cities. *Nature Climate Change*, **3** (**9**), 802–806.
2 Strauss, B.H., Ziemlinski, R., Weiss, J.L. & Overpeck, J.T. (2012) Tidally adjusted estimates of topographic vulnerability to sea level rise and flooding for the contiguous United States. *Environmental Research Letters*, **7**, 1–12.

3 City of New York (2013) *New York City Special Initiative on Rebuilding and Resiliency: A Stronger, More Resilient New York*. New York, NY.

4 Blake, R., Grimm, A., Ichinose, T. *et al.* (2011) Urban climate: processes, trends, and projections. In: Rosenzweig, C., Solecki, W.D., Hammer, S.A. & Mehrotra, S. (eds), *Climate Change and Cities: First Assessment Report of the Urban Climate Change Research Network*. Cambridge University Press, Cambridge, UK, pp. 43–81.

5 Rosenzweig, C. & Solecki, W. (eds) (2001) *Climate Change and a Global City: The Potential Consequences of Climate Variability and Change, Metro East Coast* Report for the U.S. Global Change Research Program. Columbia Earth Institute, New York, NY.

6 Rosenzweig, C., Major, D.C., Demong, K. *et al.* (2007) Managing climate change risks in New York City's water system: assessment and adaptation planning. *Mitigation and Adaption Strategies for Global Change*, **12** (**8**), 1391–1409.

7 New York City Department of Environmental Protection (NYCDEP) (2008) *Climate Change Program, Assessment and Action Plan*. New York City Department of Environmental Protection, New York, NY.

8 Metropolitan Transportation Authority (2009) *Greening Mass Transit and Metropolitan Regions: The Final Report of the Blue Ribbon Commission on Sustainability and the MTA*, New York, NY.

9 City of New York (2007) *PlaNYC: A Greener, Greater New York*, New York, NY.

10 Horton, R., Rosenzweig, C., Gornitz, V. *et al.* (2010) Climate risk information. In: Rosenzweig, C. & Solecki, W. (eds), *Climate Change Adaptation in New York City: Building a Risk Management Response*. Vol. **1196**. Blackwell Publishing on behalf of the New York Academy of Sciences, Boston, MA, pp. 148–228.

11 Major, D.C. & O'Grady, M. (2010) Adaptation assessment guidebook. In: Rosenzweig, C. & Solecki, W. (eds), *Climate Change Adaptation in New York City: Building a Risk Management Response*. Vol. **1196**. Blackwell Publishing on behalf of the New York Academy of Sciences, Boston, MA, pp. 229–292.

12 NOAA National Weather Service (2014) *Natural Hazard Statistics*. URL http://www.nws.noaa .gov/om/hazstats.shtml [accessed on 4 March 2015].

13 Gaffin, S.R., Rosenzweig, C., Khanbilvardi, R. *et al.* (2008) Variations in New York City's urban heat island strength over time and space. *Theoretical and Applied Climatology*, **94** (**1, 2**), 1–11.

14 Rosenzweig, C., Solecki, W.D., Cox, J. *et al.* (2009) Mitigating New York City's heat island: integrating stakeholder perspectives and scientific evaluation. *Bulletin of the American Meteorological Society*, **90** (**9**), 1297–1312.

15 Li, T., Horton, R.M. & Kinney, P.L. (2013) Projections of seasonal patterns in temperature-related deaths for Manhattan, New York. *Nature Climate Change*, **3** (**8**), 717–721.

16 Horton, R.M., Gornitz, V., Bader, D.A. *et al.* (2011) Climate hazard assessment for stakeholder adaptation planning in New York City. *Journal of Applied Meteorology and Climatology*, **50** (**11**), 2247–2266.

17 Horton, R., Yohe, G., Easterling, W. *et al.* (2014) Chapter 16: Northeast. In: Melillo, J.M., Richmond, T.C. & Yohe, G.W. (eds), *Climate Change Impacts in the United States: The Third National Climate Assessment*. U.S. Global Change Research Program, Washington, DC, pp. 371–395.

18 Metropolitan Transportation Authority (MTA) (2007) *August 8, 2007 Storm Report*, New York, NY.

19 The City of New York Department of Sanitation (DSNY) (2013) *2013-2014 Winter Snow Plan for the Borough of Manhattan. Pursuant to Local Law 28 of 2011*, New York, NY.

20 Meehl, G.A., Stocker, T.F., Collins, W.D. *et al.* (eds) (2007) *Climate Change 2007: The Physical Science Basis Contribution of Working Group I to the Fourth Assessment Report of the Intergovernmental Panel on Climate Change*. Cambridge University Press, Cambridge, UK, pp. 747–845.

21 Nakicenovic, N. & Swart, E.R. (eds) (2000) *Special Report on Emissions Scenarios: A Special Report of Working Group III of the Intergovernmental Panel on Climate Change.* Cambridge University Press, Cambridge, UK.

22 Pfeffer, W.T., Harper, J.T. & O'Neel, S. (2008) Kinematic constraints on glacier contributions to 21st-century sea-level rise. *Science,* **321** (**5894**), 1340–1343.

23 Horton, R., Herweijer, C., Rosenzweig, C. *et al.* (2008) Sea level rise projections for current generation CGCMs based on the semi-empirical method. *Geophysical Research Letters,* **35** (**2**), L02715.

24 Rosenzweig, C. & Solecki, W. (2010) *Climate Change Adaptation in New York City: Building a Risk Management Response.* Blackwell Publishing on behalf of the New York Academy of Sciences, Boston, MA.

25 PlaNYC (2007) *A Greener, Greater New York.*

26 City of New York (2012) *Local Laws of the City of New York for the Year 2012,* 42 pp.

27 Jacob, K., Deodatis, G., Atlas, J. *et al.* (2011) Transportation. In: Rosenzweig, C., Solecki, W., DeGaetano, A., *et al.* (eds), *Responding to Climate Change in New York State: The ClimAID Integrated Assessment for Effective Climate Change Adaptation in New York State.* Vol. **1244**. Blackwell Publishing on behalf of the New York Academy of Sciences, Boston, MA, pp. 299–362.

28 Rosenzweig, C. & Solecki, W. (2014) Hurricane Sandy and adaptation pathways in New York: lessons from a first-responder city. *Global Environmental Change,* **28**, 395–408.

29 Orton, P., Georgas, N., Blumberg, A. & Pullen, J. (2012) Detailed modeling of recent severe storm tides in estuaries of the New York City region. *Journal of Geophysical Research: Oceans,* **117** (**C9**), C09030.

30 Liu, J., Curry, J.A., Wang, H. *et al.* (2012) Impact of declining Arctic sea ice on winter snowfall. *Proceedings of the National Academy of Sciences,* **109** (**11**), 4074–4079.

31 PlaNYC (2014) *Progress Report 2014: A Greener, Greater New York; A Stronger More Resilient New York.*

32 NPCC (2015) Building the Knowledge Base for Climate Resiliency: *New York City Panel on Climate Change 2015 Report.* Rosenzweig, C. & Solecki, W. (eds.) Annals of the New York Academy of Sciences, 1336, 1–149, ISSN 0077-8923.

33 Parris, A. (2014) How Hurricane Sandy tamed the bureaucracy. *Issues in Science and Technology,* **30** (**4**), 83–90.

34 Parris, A., Bromirski, P., Burkett, V. *et al.* (2012) *Global Sea Level Rise Scenarios for the United States National Climate Assessment.* US Department of Commerce, National Oceanic and Atmospheric Administration, Oceanic and Atmospheric Research, Climate Program Office, Washington, DC.

35 State of New York Public Service Commission (2014) *PSC Approves New Rates for Con Edison.* URL https://www3.dps.ny.gov/pscweb/WebFileRoom.nsf/Web/1568A1615C240C5A85257C85006215DA/$File/pr14013.pdf?OpenElement [accessed on 4 March 2015].

36 Liu, J., Song, M., Horton, R.M. & Hu, Y. (2013) Reducing spread in climate model projections of a September ice-free Arctic. *Proceedings of the National Academy of Sciences,* **110** (**31**), 12571–12576.

37 Smith, J.B., Vogel, J.M. & Cromwell, J.E. III, (2009) An architecture for government action on adaptation to climate change. an editorial comment. *Climatic Change,* **95**, 53–61.

38 Anandhi, A., Frei, A., Pierson, D.C. *et al.* (2011) Examination of change factor methodologies for climate change impact assessment. *Water Resources Research,* **47** (**3**), W03501.

SECTION II

Managing knowledge-to-action networks

The second *Climate in Context* theme, Managing Knowledge-to-Action Networks, examines the ways in which RISA teams convene and facilitate dialogues between researchers and decision-makers. The goal of this work is to gain a shared understanding of climate-related issues and factors that influence decisions and build capacity for the use of climate information to inform resource management decisions. Managing Knowledge-to-Action Networks is built upon understanding context and risk by fostering relationships between researchers and practitioners. Within these interpersonal networks ("knowledge to action networks"), scientists can disseminate information in multiple directions—not just from the producer to the consumer of scientific information. However, it is distinct in that, apart from understanding, scientists have to manage the myriad ways contextual drivers affect relationships. Working the knowledge-to-action landscape requires the identification of the early adopters of new knowledge, the innovators and opinion leaders in a community, and the means to both facilitate communication across jurisdictions and scales of governance to overcome resistance to change. If managed well, knowledge-to-action networks sustain trust, generate insights, inform action, and spawn enduring communities of practice and learning.

The first chapter in this theme, by Stevenson *et al.*, reflects on relationships of mutual benefit between Cooperative Extension and RISA and demonstrates, through brief case studies, examples of RISA science communication, translation, and mediation in knowledge-to-action networks. The second chapter, by Olabisi *et al.*, homes in on participatory model building as a mode of catalyzing knowledge-to-action networks. Their approach is particularly effective in addressing the "wicked" or "messy" problems, which

Climate in Context: Science and Society Partnering for Adaptation, First Edition.
Edited by Adam S. Parris, Gregg M. Garfin, Kirstin Dow, Ryan Meyer, and Sarah L. Close.
© 2016 John Wiley & Sons, Ltd. Published 2016 by John Wiley & Sons, Ltd.

occur when the science is complex and the participants voice diverse values about the problem. The authors note that participatory model building empowers decision-makers to test hypotheses and generate a common scientific evidence base to inform decisions. The final chapter in Managing Knowledge-to-Action Networks, by Trainor *et al.*, focuses on knowledge sharing through webinars, given the diverse economic sectors and cultural norms, and the enormous expanse of the state of Alaska. They point out the ways in which webinar communication promotes cross-jurisdictional and cross-scale interactions (e.g., local, state, national) and provides a neutral platform for participants to explore differences with respect to perspectives as well as experience.

CHAPTER 4

Connecting climate information with practical uses: Extension and the NOAA RISA program

John Stevenson[1], Michael Crimmins[2], Jessica Whitehead[3], Julie Brugger[4] and Clyde Fraisse[5]

[1] *Climate Impacts Research Consortium, Oregon Sea Grant, Oregon State University, Corvallis, OR 97331-5503, USA*

[2] *Climate Assessment for the Southwest, Department of Soil Water and Environmental Science, University of Arizona, Tucson, AZ 85721-0038, USA*

[3] *Carolinas Integrated Sciences and Assessments, North Carolina State University, North Carolina Sea Grant, Raleigh, NC 27695-8605, USA*

[4] *Institute of the Environment, University of Arizona, Tucson, AZ 85721-0137, USA*

[5] *Southeast Climate Consortium, Department of Agricultural and Biological Engineering, University of Florida, Gainesville, FL 32611-0570, USA*

4.1 Introduction

The National Oceanic and Atmospheric Administration (NOAA) Regional Integrated Sciences and Assessments (RISA) program and Land Grant and Sea Grant incarnations of university extension systems represent two boundary organizations with unique, but complementary, structures and a capacity for connecting and mediating the boundary between climate information producers and users. The RISA program began in the 1990s to improve the link between climate science and society by supporting university consortia that could be responsive to regional stakeholder needs. Today, the RISA program funds 11 projects throughout the United States. The mission of connecting science to society is also embodied in the Cooperative Extension System established at Land Grant Universities across the United States over a century ago to produce usable science to support agricultural and resource management activities. Since then, Extension has pioneered and developed a rich history of bridging the gap between research from universities and the needs of practitioners such as agricultural producers, forest landowners,

Climate in Context: Science and Society Partnering for Adaptation, First Edition.
Edited by Adam S. Parris, Gregg M. Garfin, Kirstin Dow, Ryan Meyer, and Sarah L. Close.

and resource managers. The success of this institutional approach led to the creation of Sea Grant colleges and extension professionals focused initially on supporting fisheries management with cutting edge science. This mission has grown to cover additional areas of need including coastal hazards planning and climate change adaptation.

The similar missions of Extension and RISAs—and opportunities for their collaboration—have not been overlooked. To date, four RISAs have Extension personnel on their program staff. These connections to support both Extension and RISA functions, and provide a basis for describing how RISA-Extension collaborations have built KANs in many cases to inform the use of science in climate adaptation. Using these collaborations as the basis of case studies, this chapter reviews the relationship between these institutions and how it has supported four critical functions found in boundary organization literature: *communication*—connecting with diverse audiences; *translation*—translating between science producers and consumers; *mediation*—negotiating between science producers and consumers; and *convening*—bringing together different parties. The chapter closes with a discussion about what has worked well in the RISA-Extension partnership and how the partnership has helped build KANs, and also highlights opportunities for expanding this work.

4.2   History of Cooperative Extension

The roots of the Cooperative Extension Service, or Extension, go back to the beginning of the nation's history when agricultural societies and clubs were formed to improve agricultural techniques and disseminate agricultural information among a population overwhelmingly rural and employed in agriculture. However, Extension itself was established only in 1914, as the third component of the land-grant university system in the United States, and incorporates education, research, and Extension activities. The 1862 Morrill Act laid the foundation for this system by granting federal public lands to each state to fund land-grant colleges to provide education in agriculture and the practical arts. It was motivated by the Jeffersonian ideal of the yeoman farmer—the respectable, knowledgeable, independent, and public-spirited individual upon whose civic virtue political democracy depended—and aimed to extend access to higher education to rural and lower- and middle-class Americans who previously had little access. In addition, the Morrill Act created a sense of societal obligation for land-grant universities that was unique among national systems of higher education.

In 1877, the Hatch Act[1] added the second building block by providing funding for agricultural experiment stations, under the direction of the land-grant colleges, whose purpose was to carry out research that could be incorporated into the curriculum. It became apparent that the benefits of education and research were not actually reaching farmers, because most students were not studying agriculture, even though they were from farm families, and research did not reach farmers because they were not in college but on the farm. To address this, the Smith-Lever Act of 1914 established the Cooperative Extension Service as a partnership between the land-grant colleges and the U.S. Department of Agriculture (USDA), "to aid in diffusing among the people of the United States useful and practical information relating to agriculture … and to encourage the application of the same" (quoted in [1]). These extension services institutionalized the public service function of the land-grant institutions. The main mechanism for reaching farmers was to send agents affiliated with the land-grant institution into rural areas, where they could meet with individual farmers and hold educational activities and public demonstrations of new practices in rural areas. Today, there are Extension offices and staff in a majority of the approximately 3000 U.S. counties. This provides Extension with the institutional capacity to address issues across a broad geographical scale, in ways that are attuned to local conditions.

Evidence of the success of this system, which integrated campus-based research with the realities of agricultural production, has been provided by the rise in the productivity of American agriculture between 1920 and 1960. During this time, U.S. farmers came to successfully compete with producers anywhere in the world and agriculture was one of the most productive sectors of the economy. In addition, colleges of agriculture in land-grant universities interpreted their educational mandate to include a broad array of subjects that were pertinent to the problems of individuals, households, businesses, and governments. As further evidence of the success of this system, during this period Extension was judged to be the most trusted source of new knowledge for ordinary Americans [2].

The name "Cooperative Extension" refers to the cooperative funding arrangements for the organization by federal, state, and county governments. It has benefited from the public mandate of land-grant universities by being able to incorporate diverse and mutually reinforcing funding

---

[1]In 1890, the second Hatch Act authorized funding for land-grant colleges for African Americans; but it was not until 1994 that the Equity in Educational Land Grant Status Act initiated federal funding for existing colleges controlled by federally recognized Native American tribes.

mechanisms from these and other sources. The Smith-Lever Act of 1914 mandated that the USDA provide each state with funds based on a population-related formula. Today, these funds are distributed by the National Institute of Food and Agriculture (NIFA) within the USDA. States provide matching and additional funds for research and Extension activities, and most counties also provide support. In some states, advisory boards oversee Extension activities in their county and advise county officials on funding decisions, reinforcing the organization's accountability to local clientele. However, support from all sources has been waning in recent years, which is compromising Extension's ability to both serve its clientele as it has in the past and to respond to emerging societal and environmental issues.

4.3   Cooperative Extension as a boundary organization

Thus, from its inception, Extension has satisfied many of the organizational characteristics of a boundary organization—that is, an organization that performed several well-agreed-upon boundary managing functions to bridge science and society: convening, communication, translation, and mediation [3,4]. Initially, it was designed to bridge the boundaries between (1) land-grant universities and agricultural experiment stations, where scientists are located and research is conducted, and (2) land-grant universities and the broader public. The placement was designed to make science accessible and useful to those situated away from university centers. However, today, Extension also bridges boundaries between disciplines outside of agriculture, different levels of an organization, public and private sectors, and between developed and developing countries. All four boundary managing functions that emerge as activities such as workshop development and creation of science interpretation and synthesis products (e.g., "fact sheets"), and those that serve in science advisory roles in working groups and ad hoc committees are regularly employed by Extension professionals.

Structurally, Extension professionals assume one of two specific types of positions at the boundary of science and society. Extension "agents" are typically county-based university personnel that staff county offices and regularly interact with the local communities in the region to identify information needs and work to deliver science-based solutions to local issues. The second type of Extension professional is the "specialist" who often is required to have an advanced degree in an area of specialization (e.g., agronomy, soil science). Specialists are often based on campus, but have the responsibility to support Extension agents across the state or region. Together, agents and specialists work either to identify research needs and deliver targeted solutions with the existing information in documented studies (e.g., the communication and/or translation of existing scientific information) or involve themselves in the development of new applied research studies, which can involve other non-Extension university or agency scientists (an example of convening and mediation). This nonhierarchical network structure and diverse expertise greatly facilitates a two-way communication between and among a broad cross section of local clientele and university scientists. Intermediaries located at local institutions and at the university help limit institutional and geographic barriers that may impede the flow of information and ideas [5].

By placing agents from the land-grant institution in local counties, Extension agents develop personal relationships and place-based attachments and knowledge. Consequently, agents develop responsibility and accountability to their local constituency and advocate their perspectives and needs to the institution. This mechanism also provides Extension with the capacity to generate a deep knowledge of the local context, which is necessary for climate adaptation planning (see Part I). This knowledge can make it possible to understand how to address adaptation in a locale where climate change is controversial.

4.4   NOAA and Extension: Sea Grant

In 1963 at a meeting of the American Fisheries Society, Athelstan Spilhaus invoked the success of Land Grant colleges when he proposed establishing Sea Grant colleges "to promote the relationship between academic, state, federal, and industrial institutions in fisheries" [6]. Just over 100 years after the Land Grant University System was established, the U.S. Congress passed the National Sea Grant College Program Act of 1966 to establish a similar university-based system in every coastal U.S. state and territory that focused on coastal and marine issues [7,8]. The administrative responsibility to initiate and support marine advisory programs that encompassed education, research, and Extension by Sea Grant colleges was originally assigned to the National Science Foundation [7]. Sea Grant was transferred to the NOAA in 1970, and under NOAA has grown to include 33 Sea Grant programs in every coastal and Great Lakes state, including Guam and Puerto Rico, as well as the National Sea Grant Library and the National Sea Grant Law Center.

Sea Grant program structure may vary from state to state, but all programs support an active suite of integrated research, outreach, and education programs that provide tangible benefits for the communities and industries in their states [9]. NOAA provides funding to state programs every two years, and each program submits an omnibus proposal that includes a portfolio of research projects as well as funding for communications, education, and program management [7]. Collectively, the Sea Grant programs deliver Extension programming through the efforts of over 400 Extension professionals [7,8]. While managed separately from the Land Grant and Cooperative Extension Service, there are still numerous ties between Sea Grant and its land-based inspirations. Three of the first four Sea Grant programs that were designated as receiving Sea Grant College Program status in 1971 were located at Land Grant institutions [6], and about two-thirds of Sea Grant Extension programs are linked to their home state's Extension program [7]. As with Extension's expansion beyond agricultural issues, Sea Grant's focus has expanded beyond its original fisheries roots to include many coastal sectors. Sea Grant researchers and Extension professionals use both physical and social science to inform decisions on many coastal issues, with special national efforts in the areas of healthy coastal ecosystems, sustainable coastal development, a safe and sustainable seafood supply, and hazard resilience in coastal communities [9].

4.5   Boundary management: RISAs and Extension working together

Today, scholars describe much of the effort made by Extension and the RISAs to link science and information with practice as *boundary management*. Research on information delivery has shown that successful production and transfer of technical information greatly benefits from the presence of organizations that manage or bridge the boundary between different communities who do not share the same norms, language, or expectations, particularly around what may constitute usable information [3]. In the case of climate adaptation, this boundary can exist as the space between the science of climate and climate impacts (e.g., hydrology, forestry, agronomy) and the practice of adaptation planning. Boundary organizations such as Extension and the RISAs manage the interface among these groups who may come to the table with vastly different experiences when it comes to developing science, and applying it to adaptation planning.

The intellectual roots describing institutional boundaries are somewhat counterintuitive to the discussion above; however, drawing on Gieryn [10] who describes boundary *work* not as a breaking down of divisions but rather as the *construction* of boundaries to maintain the integrity of one's institutional practice. For example, defending its understanding of the natural world over religious explanations, scientists of the 19th century had to clearly demarcate between what constituted scientific and nonscientific [10]. Similarly, policy makers may resist the inclusion of scientists for fear of what Guston [11] refers to as the "scientization of politics," or, disguising normative values as scientific facts [11]. Difficult as it may be, there is evidence that modern day boundary *managers*—referred to here as those bridging boundaries—can lead to improved action, particularly around problems related to climate change and adaptation planning [12]. It is evident from this literature that Extension and RISAs are not simply managing institutional differences, but differences that may be actively reinforced by the parties that boundary managers are trying to bring to a common table.

To address these barriers, Cash *et al.* [3] identified four roles particularly critical to boundary management. These include communication, translation, mediation, and convening. *Communication* is two way, occurring from experts to users and users to experts, iterative, and inclusive to a wide range of perspectives. *Translation* is the interpretation between communities of practice (e.g., famers and university researchers), or place (e.g., coastal residents and noncoastal residents), which may understand the same things but describe them differently. For example, a boundary manager is in the unique position to know that words like "central tendency" used by experts can also mean "average" to a nonexpert or interpret local names of geography or species to nonlocals. *Mediation* refers to managing a process in which differing communities interact. In particular, boundary workers support these collaborations by ensuring that efforts are transparent, accepting of differing perspectives, or helping to establish and enforce behavioral guidelines. In other words, making sure everyone plays nicely [3]. A fourth function in boundary management is *convening* or the bringing together of these parties [4].

As suggested above, boundary management has long been a function of Extension professionals and the RISAs. Our review of boundary management related to climate adaptation shows Extension and RISAs filling a similar role. Some of this work has occurred through educational programs via communication and translation functions, while other efforts have focused on the development of innovative stakeholder and researcher collaborations by providing mediating and convening functions. The goal of these collaborations, sometimes referred to as "knowledge action systems" or "knowledge-to-action networks" (KANs), is to provide a venue to link technical information to decision making through the "coproduction" of knowledge around climate impacts and adaptation [13]. In the following case studies, we review how Extension and RISA collaborations supported adaptation planning through these boundary management functions.

4.5.1 Climate Impacts Research Consortium

The Climate Impacts Research Consortium (CIRC) began work as the Pacific Northwest RISA in 2010. In its proposal to NOAA, CIRC set out to support landscape and watershed management in a changing climate. To support this work, CIRC convened a regional team of interdisciplinary experts in climatology, hydrology, forestry, coastal geology, and sociology from six universities in Idaho, Oregon, Washington, and Western Montana.[2] CIRC also proposed a partnership with Extension from each state to support a full time Regional Extension Climate Specialist (RECS) to serve as a link between the research consortium and the region's network of Extension professionals and communities.

CIRC's approach to adaptation support is based on six principles identified by the National Research Council: "(1) begin with user needs; (2) give priority to process over products; (3) link information producers and users; (4) build connections across disciplines and organizations; (5) seek institutional stability; and (6) design processes for learning" [14]. From an Extension perspective, CIRC has pursued this approach through two efforts. The first is to develop and maintain KANs, through which CIRC works with information users to coproduce information to address management problems. The second effort is an educational program aimed at conveying these findings and building climate literacy in the regional community.

During its first year of funding, CIRC engaged regional stakeholders with a science needs assessment survey directed toward managers and resource users across Idaho, Oregon, and Washington. In the spring of 2011, CIRC organized a series of workshops to meet with stakeholders face to face to discuss management challenges and relevant climate science that could be brought to bear on these issues. The results from these assessments led to a research agenda focused on developing integrated scenarios of future climate, hydrology, and vegetation in the Pacific Northwest with particular emphasis on providing decision support for a wide range of management issues. The assessment also resulted in CIRC assembling additional expertise in economics and public health, which stakeholders identified as topics to consider in impacts research.

Extension has helped carry this agenda forward through development of KANs via the RECS and local Extension networks. One of CIRC's early KANs is concerned with climate impacts on water supply with participation from local water users, land use planners, nongovernmental organizations, as well as local, state, and federal managers from the Big Wood River

---

[2]Boise State University, University of Idaho, Oregon State University, University of Oregon, University of Washington, and Montana State University.

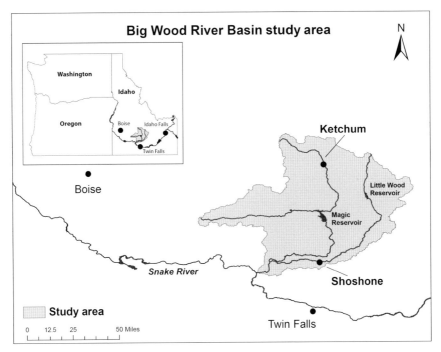

Figure 4.1 Map of the Big Wood River Basin study area.

Basin (Figure 4.1) in central Idaho. The project originated during follow-up conversations during the spring regional workshops about needed projections for water supply for the Snake River in southern Idaho. In the following summer, the RECS toured along the Snake River from western Idaho to its headwaters on the western slope of the Teton Mountain Range. During that tour, the RECS worked through Idaho Extension's community networks to meet with local water users and managers to better understand current water issues and discuss how future climate conditions may influence these and other management challenges. Working with this initial list of contacts, the CIRC researcher and Extension team, including local extension agents, continued discussions that led to a pilot project in the Big Wood Basin to explore how climate and other drivers, such as population growth and land use change, would influence future water supplies for municipal and irrigation uses. Through in-person workshops and webinars, CIRC has collaborated with this network to coproduce model representations of physical, biologic, and human processes in the basin and to explore how they will respond to future climate change under alternative management scenarios [15]. This project has relied heavily on the RECS and the CIRC team, which implicitly bring Extension and RISA together as collaborators, to provide boundary services, for example, convening and mediating between researchers and stakeholder participants during workshops, and translating technical information such as trade-offs between different modeling approaches.

More recently, CIRC has begun working with a KAN focused on the response of coastal flooding and erosion hazards to sea level rise along the Oregon Coast. As of this writing, this project is in its early stages but has developed based on an existing network of property owners and local and state agency staff working with Oregon Sea Grant Extension and expanding it to a wider geography and set of stakeholders. Similar to the work on the Big Wood Basin project, Sea Grant Extension has facilitated these networks by managing the boundaries between local stakeholders, agency staff, and university researchers.

Extension has also supported CIRC's role in bringing an understanding of regional climate and adaptation science to the community level through education activities. In early 2011, the RECS and CIRC PIs worked with Extension leadership from each state to deliver in-service trainings to regional Extension professionals. The program provided an overview of the climate in the Pacific Northwest, model projections for future change, and research that looked at how critical resources such as water, vegetation, and cropping systems may be impacted over the next 50–100 years. These workshops also led to conversations about additional products CIRC could provide to support climate education. As a result of those workshops and other stakeholder conversations, CIRC now produces the *CIRCulator*, a monthly digest of newly published research on climate for the region, and more recently, the *PNW Climate Recap and Outlook*, which provides a recap of past seasonal trends and an outlook for the subsequent 3 months. CIRC has also developed educational material for existing Extension programs in forestry, agriculture, and Sea Grant via public presentations, content for training modules, and webinars.

4.5.2 Southeast Climate Consortium

The Southeast Climate Consortium (SECC) RISA is a consortium of six universities[3] from Florida, Georgia, and Alabama. SECC's mission is to use advances in climate science and practice to provide scientifically sound information and decision-support tools for agro-ecosystems, forests, and other terrestrial and coastal ecosystems of the southeastern United States. This work was supported by SECC convening expertise from personnel from physical, biological, and social science disciplines.

---

[3]Florida State University, University of Florida, University of Miami, University of Georgia, Auburn University, and University of Alabama at Huntsville.

Early in 2004, SECC and regional Extension began collaborating to deliver climate educational programs and together they hired the country's first climate Extension specialist at the University of Florida. This decision was based on a number of important aspects of Extension including (1) an established network of Extension faculty at the county level; (2) a mission to provide unbiased, research-based answers that are relevant to stakeholders; and (3) experience in delivering educational programs using effective language and formats. SECC's vision is to improve agricultural, forestry, and water resource managers' capacity to cope with uncertainty and climate-associated risks through effective use of climate information. This vision includes a major role for Extension through an outreach and education program [16]. Many of the SECC and regional Extension partnerships represent KANs in which Extension helps facilitate relationships with regional stakeholders (Table 4.1) and reflect the core functions of boundary work from communicating and translating information to mediating relationships.

SECC's overall goal is to prepare stakeholders to face challenges posed by climate variability and change by increasing relevant knowledge of citizens, professionals, and agency personnel, as well as developing collaborative solutions and actions that can reduce potential climate impacts on the natural and built environments. Decision-support systems are a key aspect in this process, which SECC and Extension collaboratively pursued primarily through the development of *AgroClimate*[4] [26]. *AgroClimate* provides climate forecasts combined with risk management tools for a range of crops, forestry, pasture, and livestock. The system was developed to allow easy expansion of topic areas, number of commodities, and risk management tools available for users at different locations in the Southeast. Participatory approaches were used for research, development, and dissemination of *AgroClimate* [21,27]. Extension was a catalyst for SECC's work on understanding the context of producers through the use of sondeos (a semi-structured, multidisciplinary, team-discussion process), focus groups, semi-structured interviews, web-surveys, on-line feedback, and participation in farmer association meetings. Thus, boundary work (i.e., convening, communicating, translating, and mediating) was a fundamental part of the *AgroClimate* development process. Prototypes were developed and used for feedback before a final version of the system was implemented.

More recently, SECC has developed alternative mechanisms to engage agricultural producers in a discussion about climate variability and change. Under a project funded by the USDA, the NIFA Climate Change program, SECC developed annual "Climate Adaptation Exchange Fairs" to engage

---

[4]http://www.agroclimate.org.

Table 4.1 Example activities that demonstrate knowledge–action-networks within SECC.

Knowledge–Action networks	Substantive advance	Action	Source
Florida Cooperative Extension Service partnership— agriculture and ranching	Establishment of trust and infrastructure planning for information dissemination	Initial discussions for the implementation of a climate information system in Florida	[17]
Water resources	Identification of nursery/landscaping, crop associations, water management districts, and water utilities as a distinct network for seasonal forecast applications	Extension Agents use climate forecasts for education	[18]
Forestry, wildfire, social-economic	Development of distinct and inter-related climate extension to multiple communities	Development of AgroClimate decision support tool and specialized wildfire, freeze, and crop-specific yield forecasts	[16,19]
Alabama and Georgia Extension Services partnerships	Network expansion	Workshops, in-service training, and development of interpreted materials preferred by extension agents	[20]
Insurance companies, banks, urban planners, marine management, tourism	Refinement of decision-support tools and methods, and development of SECC as a multistate institution	Fine spatial resolution information delivered, in response to user requests	[21]
Cooperative Extension Service partnership— agriculture and ranching	Knowledge coproduction	In-service training of extension personnel, Extension Agents and crop growers increasingly find climate forecasts information helpful	[22,23]

Table 4.1 (*continued*)

Knowledge–Action networks	Substantive advance	Action	Source
Georgia farmers	Demonstration of the social nature of risk management for the integration of climate predictions into agricultural management practices	Identification of crop selection and planting timing as key decisions influenced by climate	[23]
North Carolina Cooperative Extension Service	Increased understanding of extension and farmers' needs in terms of decision tools and climate information	Baseline survey of North Carolina Cooperative Extension agents	[24]
Agriculture-based Climate Working Groups	Purposeful participatory research-to-action engagement process; knowledge exchange	Increased Extension Agent use of AgroClimate; grower focus on seasonal change	[25]

producers, Extension, and researchers in discussions on how specific technologies and management alternatives can reduce climate-related risks (adaptation) and increase resource-use efficiency (mitigation). A typical event might include a panel of producers discussing their use of experimental technologies or management strategies and describing the benefits and barriers of the featured technologies. This event has reinforced SECC's assumption that risk reduction strategies generated by the agricultural community are some of the most promising options for adapting to climate change [28]. At the time of this writing (2014), SECC has organized two regional exchange fairs that improved producers' engagement with the idea of preparing for a changing climate in Southeast United States.

4.5.3 The Carolinas Coastal Climate Outreach Initiative: a South Carolina Sea Grant Consortium, North Carolina Sea Grant, and CISA partnership

Sea Grant's consideration of climate issues in coastal areas began in the early 1990s, but the Sea Grant-RISA collaboration on climate began in the mid-2000s [29]. In 2006, the National Sea Grant Office and NOAA Climate Program Office provided joint funding for a five-year pilot to establish a coastal climate Extension program. The funding was awarded to a three-way partnership between the Carolinas Integrated Sciences and Assessments RISA (CISA) and state Sea Grant programs in North Carolina and South Carolina. As part of this funding, the partnership hired a Regional Climate Extension Specialist (RCES); the first Extension specialist in the Sea Grant network to be devoted solely to climate variability and climate change impacts on coastal regions. Funding for the program continued beyond 2012 when the RCES became a funded CISA coprincipal investigator, with the remainder of the position's funding provided by SC Sea Grant and NC Sea Grant. By 2013, demand for Extension programming around coastal climate activities justified maintaining a Sea Grant Climate Extension Specialist solely focused in South Carolina and establishing a new Hazards Adaptation Specialist position in North Carolina.

The Extension programs and activities conducted through the CISA-Sea Grant Extension partnership (CISA-Extension) fulfilled four general objectives: (1) develop the capacity of the state Sea Grant Extension programs to inform and educate coastal decision-makers of the implications of climate variability and change for major coastal issues; (2) provide tailored, decision-relevant information on the implications of climate variability and change to coastal decision-makers; (3) increase the capacity of the Sea Grant network regionally and nationally to deliver outreach programs and conduct relevant research on the impacts of climate variability and change on coastal stakeholders; and (4) evaluate and review enhancements in Sea Grant Extension and education capacity and approaches. The partnering institutions functioned as a team led by the RCES, with CISA providing scientific input on climate variability, climate change, and adaptation, and the two Sea Grant Extension programs contributing their knowledge of local stakeholder networks and decision needs. The team used an informal needs assessment of selected key stakeholders suggested by other Sea Grant Extension professionals to develop programs that would fulfill key needs for communication, translation, and mediation. Additionally, efforts to convene Extension and outreach expertise both in the region and nationally, in partnership with other organizations, gave the partnership a key role in convening communities of practice that expanded Sea Grant's climate outreach capacity.

Early in the CISA-Extension partnership it became apparent that attempts to communicate climate information that were divorced from the local, decision-relevant contexts would be ineffective (e.g., [30–32]). To address this challenge, the CISA-Extension partnership helped develop the Vulnerability, Consequences, and Adaptation Planning Scenarios (VCAPS) process, led by the Social and Environmental Research Institute (SERI) (see Chapter 1). VCAPS is a facilitated process that uses the causal structure of hazards and climate vulnerability as a basis for leading groups of decision-makers through discussions of outcomes, consequences, public and private actions, and contextual factors while capturing the conversations on diagrams in real-time [33]. As with system dynamics models used by Schmitt Olabisi *et al.* (Chapter 5), the causal pathway diagrams generated during the VCAPS process function as boundary objects that document the coproduction of knowledge between scientists and decision-makers and reflect a common stakeholder understanding of issues [34,35]. Since its original development and use in Sullivan's Island, SC [35], SERI, CISA, and Sea Grant have facilitated VCAPS exercises in McClellanville, SC, Plymouth, NC, and Beaufort, SC; VCAPS has also been facilitated by SERI and other organizations in seven other communities outside the Carolinas.

VCAPS has been a particularly effective decision-support process for the CISA-Extension partnership because of its simultaneous support for communication and mediation among decision-makers and translation of scientific information to decision-making contexts. During the Carolinas VCAPS processes, the RCES provided climate expertise as well as facilitation, combining the communication and mediation functions to ensure that iterative two-way communication about the science took place throughout the diagramming exercises. However, part of the VCAPS success stems from the facilitation team creating an environment in which participants assume some degree of responsibility for the communication, translation, and mediation roles, increasing stakeholder buy-in for the results [33]. In the Carolinas, initial climate and hazards presentations included as part of the VCAPS process focused on observed data and the broad range of future scenarios. Participants then discussed the potential consequences of the data on the community and generated ideas for actions that could be adapted by both public and private entities, thereby taking on the role of translating science into understanding and action themselves. This facilitated the process that allowed for mediation among many different stakeholders during the diagramming exercise by providing decision-makers an opportunity to discuss climate-related hazards across often stove-piped departments. For example, in follow-up interviews a Sullivan's Island planning commission member noted that "(w)e rarely get a chance to sit down and talk and pick (town staff specialists') brains" about wastewater treatment issues [35]. In some cases, VCAPS also served as a starting point for participating decision-makers to serve as translators to engage other local stakeholders. For example, the VCAPS process in Plymouth, NC, led to additional discussions between the town and a local NGO about potential inundation during hurricanes [33].

The CISA-Extension partnership also performed the convening function by contributing to the formation of regional and national coastal climate outreach communities of practice. In 2009, CISA-Extension partners took a central role in establishing the Sea Grant Climate Network, a grassroots organization of Extension, communications, and educational professionals in the Sea Grant network who included climate information in their work but may not necessarily be climatologists or climate scientists by training. On the regional level, CISA and Sea Grant climate outreach personnel were a part of the team that used funding through the National Sea Grant Office and NOAA's Southeast and Caribbean Regional Team to conduct two regional workshops for coastal outreach professionals on climate adaptation and outreach in the Southeast and Caribbean Climate Community of Practice.[5] In 2013, the CoP hosted its first webinar, and CISA played an integral role in conducting a CoP member survey to fill the need for a regional resource directory that facilitates collaboration among members. At the time of writing, the RCES serves as the chairperson of the CoP and CISA provides logistical support. Through these networking efforts and other partnerships, the CISA-Extension partnership helped to support climate-related programming offered by groups such as the National Estuarine Research Reserves Coastal Training Programs, state Coastal Zone Management programs, the South Carolina Coastal Information Network, NOAA in the Carolinas, the Governor's South Atlantic Alliance, and others in the Carolinas.

By conducting applied projects in communities and expending efforts toward convening communities of practice, the partnership of South Carolina Sea Grant, North Carolina Sea Grant, and CISA helped to establish Sea Grant and CISA as key boundary organizations for providing coastal climate information to decision-makers in South Carolina and North Carolina [36]. The partnership has served as an example for other RISA and Sea Grant programs across the United States, many of which are now forging similar partnerships. The CISA-Extension partnership's success increased the demand for coastal climate projects in both states, leading to a program restructuring with each of the two states' Sea Grant programs employing dedicated climate and hazards Extension staff. This structure will meet stakeholder needs and preserve the partnership in the Carolinas, enhancing climate programming impacts in both states in the years to come.

4.5.4 University of Arizona and the Climate Assessment for the Southwest

A partnership between the NOAA Climate Program Office, within which the RISA Program resides, and the University of Arizona College of Agriculture and Life Sciences was forged in 2004 to develop an Extension specialist

---

[5]http://stormsmart.org/groups/sec-ccop/.

position that would work to connect federal agency—based climate science and information with decision-makers and resource managers across the southwest United States. The "Extension specialist" position, often housed on-campus within an academic unit, has the responsibility of developing an applied research and Extension program that directly supports stakeholders (including county Extension agents) by providing access to cutting edge research and information and by participating in experimental efforts (e.g., crop trials, utilizing experimental forecast information). The development of an "Extension Specialist in Climate Science" position at the University of Arizona was unique in that it explicitly linked the Land Grant University with a partner agency to leverage each other's resources and networks. This partnership was modeled after earlier and successful partnerships formed between Land Grant Universities and NASA in the development of Geospatial Extension Specialist positions at several institutions across the country.[6] Initial funding for the position was provided by the sponsoring agency, NOAA, for three years after which the specialist would be supported by the university as a full, tenure track faculty member in a host department.

The initial position description for the Extension Specialist in Climate Science made explicit linkages between the home department and the Climate Assessment for the Southwest (CLIMAS) RISA at the University of Arizona. The position was designed for the specialist to be an affiliate researcher with the CLIMAS program to help connect the Extension network of specialists, agents, and stakeholders with the network of scientific expertise developed by the RISA program. CLIMAS had strong ties with University of Arizona Cooperative Extension (UACE) prior to the development of this position, but this formalized connections between the two institutions helping to leverage resources and capacity in both directions and improving coordination on issues related to climate science research, education, and outreach.

Embedding an Extension specialist within CLIMAS has helped to strengthen and enhance the boundary organization functions of both the Extension network, by strengthening climate science, and the RISA program, by strengthening connections to diverse audiences and sectors that are sensitive to climate. Now over 12 years old, the monthly climate newsletter called the "Southwest Climate Outlook" (SWCO) is a good example of the power of Extension-RISA connections with respect to science communication and translation boundary functions. SWCO is produced by a team comprising CLIMAS researchers and Extension professionals that are embedded within the CLIMAS team. It is read by several hundred ranchers,

---

[6]http://www.geospatialextension.org/.

farmers, resource managers, and decision-makers across the Southwest each month and is actively promoted as a climate monitoring tool by Extension agents across Arizona. Lessons learned in the production of the SWCO and positive feedback on the utility of synthesizing and translating existing climate information into a "value added" product led to further experimental efforts. In 2010, an impending La Niña event was expected to bring below-average precipitation to the Southwest and worsening drought conditions. CLIMAS researchers (including two Extension professionals) teamed up to rapidly develop a drought-monitoring information product called the "La Niña Drought Tracker" (LNDT). Regular communication between CLIMAS researchers and Extension professionals, including Extension agents throughout Arizona, provided insight into the growing concern over the development of drought conditions and the potential utility of targeted communication and guidance on the developing La Niña event. The LNDT was produced monthly between December 2010 and May 2011 when a La Niña event did indeed develop and subsequent winter precipitation amounts were less than 50% for much of Arizona and New Mexico. Since the product was experimental, the development team wanted to gain some insight into the utility of this effort in supporting climate monitoring, decision-making, and outreach. Survey results indicated that over 400 people consulted the product each month and a majority of them used the information it provided to support a drought-related decision [37]. This finding supports an initial hypothesis that interpreting and synthesizing existing climate information is an important boundary organization function further strengthened through the RISA-Extension partnership existing at the University of Arizona. This project and end result was also an articulation of the "knowledge-to-action" concept with both Extension and RISA at the core of the network. Results indicate that usable information was attained through the expert interpretation in the final synthesis product.

Another example of a project that exemplifies the RISA-Extension connection in Arizona and demonstrates boundary organization functions was conducted in 2011 as part of the National Climate Assessment. A project titled "Weather, Climate and Rural Arizona: Insights and Assessment Strategies" [38] was conducted by a team consisting of an Extension specialist and a CLIMAS staff social scientist to assess attitudes, perceptions, and information needs with respect to climate change across the region. This project leveraged the Extension network across the state to gain access to "opinion leaders" across rural Arizona by working with and through local county Extension offices. County Extension directors were interviewed with questions related to local attitudes and perceptions of climate change as well as the role of Extension in delivering climate science research and information. They were also asked to compile lists of key local officials, resource managers, agriculturalists, and citizen leaders whose participation could help contribute to a constructive discussion in focus groups, a prime example of Extension's capacity in *convening* meetings of diverse groups. Meetings centered on several focus group discussions that were held in 9 of Arizona's 15 counties over a six-month period in 2011. The discussions provided key insight into the fact that many rural Arizonans are highly attuned to weather and climate patterns across the region and have been observing what they consider to be impacts and changes outside of historical norms [39]. These changes were attributed to some factors that invoked anthropogenic climate change as a probable cause, while others referred to long-term cycles. Each group that was convened included County directors and agents who participated and helped mediate heated discussions on sometimes controversial topics (e.g., climate change attribution, government regulations) and also served as communicators and translators between participants representing different sectors or locations within the county.

4.6   Conclusions

Our review of these four RISA-Extension case studies is not an exhaustive account of the partnerships' efforts but is intended to provide examples of how these two institutions have worked together to generate mutual benefits and to advance adaptation planning. In particular, the case studies demonstrate how RISAs and Extension used their respective convening power with interdisciplinary scientists and local communities to form more inclusive KANs around climate adaptation planning than either could have done independently. CIRC's water futures project in Idaho, SECC's development of the AgroClimate, CLIMAS' LNDT, and CISA's efforts around adaptation planning scenarios (VCAPS) all illustrate how the RISA-Extension partnership expands the number of parties convened, strengthens mediation during the process of engagement, and provides translation of technical and local information that arose in these processes. CLIMAS' efforts demonstrate how a RISA-Extension partnership functions to communicate and translate climate information through products such as the SWCO and LNDT.

   In many of these cases, the partnerships served a critical role in extending the reach of RISA efforts through the careful utilization of an extensive and well-trusted Extension network [5,39,40]. Yet, the RISA-Extension partnership also demonstrates new ways to develop usable information for problems as intractable as climate change. While Extension bears a long legacy in the development of applied research with stakeholder input, the KAN projects all exemplify advances in developing practical information where stakeholders are recruited and enlisted as collaborators and experts in their own right rather than clientele waiting for end products—a mode of operation embraced by the RISA program, from its inception. In this way, the partnership with the RISAs has given Extension access not just to expertise on climate science but also to the latest advances in social science in the understanding of engaging and helping communities adapt to change (see Understanding Context and Risk).

These case studies also help to reflect on future opportunities for the RISA-Extension in supporting climate adaptation. Four of the RISA projects have formal partnerships with their state Extension services and Sea Grant, which means that there are opportunities to expand the connections among remaining projects and their respective Extension institutions and Sea Grant programs. Formally connecting RISA and Extension programs through the creation of unique positions and institutional arrangements appears to enhance the level of coordination and potential leveraging of networks and resources. Embedding Extension personnel within RISA projects, which began as an experiment with SECC, is now part of the RISA "DNA"—a strategy to create strong institutional partnerships to benefit each of the collaborating organizations. Looking further ahead, there also seems to be opportunities for Extension services to expand their role in supporting climate adaptation via support from RISA projects. One of the efforts of the CIRC discussed above focused on delivering in-service training on climate science to Extension faculty in the Pacific Northwest and underscores opportunities for increasing Extension's own capacity through RISA-supported train-the-trainer programs. Doing so on a broader scale has the potential to amplify RISA investments through these long-established networks of boundary workers.

References

1 Cash, D. (2001) In order to aid in the diffusing useful and practical information: agricultural extension and boundary organizations. *Science, Technology, and Human Values*, **26** (**4**), 431–453.

2 McDowell, G.R. (2001) *Land-Grant Universities and Extension into the 21st Century: Renegotiating or Abandoning a Social Contract*. Iowa State University Press, Ames.

3 Cash, D.W., Clark, W.C., Alcock, F. *et al.* (2003) Knowledge systems for sustainable development. *Proceedings of National Academy of Sciences*, **100** (**14**), 8086–8091.

4 Cash, D.W., Borck, J.C. & Patt, A.G. (2006) Countering the loading dock approach to linking science and decision making: comparative analysis of El Niño/Southern Oscillation (ENSO) forecasting systems. *Science Technology & Human Values*, **31** (**4**), 465–494.

5 Brugger J, Crimmins M. (2014) Designing institutions to support local level climate change adaptation: insights from a case study of the US Cooperative Extension System. *Weather, Climate and Society* **7,** 18–38. doi: 10.1175/WCAS-D-13-00036.1

6 Spilhaus, A. (1972) Birth of a Sea Grant land is just an island. *Eos, Transactions American Geophysical Union*, **53**, 572–578.

7 National Sea Grant College Program. (2013) *Fundamentals of a Sea Grant Extension Program, Second Edition.* Bunting-Howarth, K., Bacon, R., Balcom, N. *et al.* (eds) *Cornell University Report 4/13.75M NYSGI-H-13-001*, pp. 96. Cornell University, Ithaca, NY. http://seagrant .noaa.gov/Portals/0/Documents/how_we_work/outreach/extension_fundamentals_web_ final-2013.pdf [accessed on 6 March 2015].

8 Susko, E., Spranger, M., Tupas, L. *et al.* (2013) *The Role of Extension in Climate Adaptation in the United States: Report from the Land Grant – Sea Grant Climate Extension Summit.* March 13–14, 2012, pp. 34.Silver Spring, MD. http://seagrant.noaa.gov/Portals/0/Documents/what_we_ do/climate/Sea%20Grant_Climate%20Extension%20Summit%20Report.pdf [accessed on 6 March 2015].

9 National Sea Grant College Program. (2014) *Strategic Plan 2014–2017: Sustaining Our Nation's Ocean, Coastal, and Great Lakes Resources Through University-Based Research, Communications, Education, Extension and Legal Programs*, pp. 28. http://seagrant.noaa.gov/Portals/ 0/Documents/global_docs/strategic_plan/2014-2017_National_Sea_Grant_College_ Program_Strategic_Plan_v2.pdf [accessed on 6 March 2015].

10 Gieryn, T. (1983) Boundary-work and the demarcation of science from non-science: strains and interests in professional ideologies of scientists. *American Sociological Review*, **48**, 781–795.

11 Guston, D. (2001) Boundary organizations in environmental policy and science. *Science, Technology, & Human Values*, **26** (**4**), 339–408.

12 Dilling, L. & Lemos, M.C. (2011) Creating usable science: opportunities and constraints for climate knowledge use and their implications for science policy. *Global Environmental Change*, **21**, 680–689.

13 National Research Council. (2005) *Knowledge–Action Systems for Seasonal to Interannual Climate Forecasting: Summary of a Workshop.* D.W. Cash & J. Buizer, rapporteurs. Roundtable on Science and Technology for Sustainability. Policy and Global Affairs. The National Academies Press, Washington, DC.

14 National Research Council (2009) *Informing Decisions in a Changing Climate.* The National Academies Press, Washington, DC.

15 Marshall, A., Lach, D., Stevenson, J. *et al.* (2016) Collaborative modeling to assess climate impacts on water resources in the Big Wood Basin, Idaho. In: *Including Stakeholders in Environmental Modeling: Considerations, Methods and Applications.* Springer Publishing.

16 Fraisse CW, Bellow J, Breuer NE *et al.* (2005) *Strategic Plan for the Southeast Climate Consortium Extension Program.* Technical Report Series 05-02. The Southeast Climate Consortium, Gainesville, FL.

17 Jagtap, S.S., Jones, J.W., Hildebrand, P. *et al.* (2002) Responding to stakeholder's demands for climate information: from research to applications in Florida. *Agricultural Systems*, **74**, 415–430.

18 Breuer, N., Canales, G., Cabrera, V. *et al.* (2005) *Potential Applications of Seasonal Climate Forecasts for Water Management and Extension Agent Perceptions of Water Issues in South Florida.* Southeast Climate Consortium Technical Report Series. SECC-05-005. Southeast Climate Consortium, Gainesville, FL.

19 Vedwan, N., Broad, K., Letson, D. *et al.* (2005) *Assessment of ClimateIinformation Dissemination by the Florida Climate Consortium.* Southeast Climate Consortium Technical Report Series: SECC-05-001. Southeast Climate Consortium, Gainesville, FL.

20 Cabrera, V., Breuer, N., Bellow, J. *et al.* (2006) *Extension Agents' Knowledge and Perceptions of Seasonal Climate Forecasts in Florida.* Southeast Climate Consortium Technical Report Series: SECC Technical Report 06-001. Southeast Climate Consortium, Gainesville, FL.

21 Breuer, N.E., Cabrera, V.E., Ingram, K.T. *et al.* (2008) AgClimate: A case study in participatory decision support system development. *Climatic Change*, **87**, 385–403.

22 Breuer, N.E., Fraisse, C.W. & Cabrera, V.E. (2010) The Cooperative Extension Service as a boundary organization for diffusion of climate forecasts: a 5-year study. *Journal of Extension*, **48**, 4RIB7.

23 Crane, T., Roncoli, C., Breuer, N.E. *et al.* (2010) Forecast Skill and Farmers' Skills: Seasonal Climate Forecasts and Agricultural Risk Management in the Southeastern United States. *Weather, Climate, and Society*, **2**, 44–59.

24 Dinon, H., Breuer, N., Boyles, R. *et al.* (2012) *North Carolina Extension Agent Awareness of and Interest in Climate Information for Agriculture.* Southeast Climate Consortium Technical Report Series: SECC Technical Report 12-003:81–124. Southeast Climate Consortium, Gainesville, FL.

25 Bartels, W., Furman, C.A., Royce, F. *et al.* (2012) *Developing a Learning Community: Lessons from a Climate Working Group for Agriculture in the Southeast USA.* Southeast Climate Consortium Technical Report Series: SECC Technical Report 12-001:1–56. Southeast Climate Consortium, Gainesville, FL.

26 Fraisse, C.W., Breuer, N.E., Zierden, D. *et al.* (2006) AgClimate: a climate forecast information system for agricultural risk management in the southeastern USA. *Computers & Electronics in Agriculture*, **53** (**1**), 13–27.

27 Breuer, N.E., Fraisse, C.W. & Hildebrand, P.E. (2009) Molding the pipeline into a loop: the participatory process of developing AgroClimate a Decision Support System for climate risk reduction in agriculture. *Journal of Service Climatology* (On-Line), **1**, 1–12.

28 Fraisse CW, Breuer NE, Zierden D et al. (2009) From climate variability to climate change: challenges and opportunities to extension. *Journal of Extension* [On-line] **47**(**2**), Article 2FEA9. Available at: http://www.joe.org/joe2009april/a9.php.

29 Sea Grant Climate Network. (2009) *Charter: Sea Grant Climate Network.* http://sgccnetwork.ning.com/page/sea-grant-climate-network [accessed on 25 July 2013].

30 Dow, K. & Carbone, G. (2007) Climate science and decision making. *Geography Compass*, **1** (**3**), 302–324.

31 Failing, L., Gregory, R. & Harstone, M. (2007) Integrating science and local knowledge in environmental risk management: a decision-focused approach. *Ecological Economics*, **64** (**1**), 47–60.

32 Shafer, M.A. (2008) Climate literacy and a National Climate Service. *Physical Geography*, **29** (**6**), 561–574.

33 Webler, T., Tuler, S., Dow, K. *et al.* (2014) Design and evaluation of a local analytic-deliberative process for climate adaptation planning. *Local Environment: The International Journal of Justice and Sustainability*, 1–23.

34 Parris, A., Simpson, C. & Abdelrahim, S. (2010) *RISA Workshop Report: Looking Ahead at Climate Service, Assessment, and Adaptation.* NOAA, Washington, DC.

35 Kettle, N., Dow, K., Tuler, S. *et al.* (2014) Integrating scientific and local knowledge to inform risk-based management approaches for climate adaptation. *Climate Risk Management*, **4–5**, 17–31. doi:10.1016/j.crm.2014.07.001

36 Tribbia, J. & Moser, S. (2008) More than information: what coastal managers need to plan for climate change. *Environmental Science and Policy*, **11** (**4**), 315–328.

37 Guido, Z., Hill, D., Crimmins, M. *et al.* (2012) Informing decisions with a climate synthesis product: implications for regional climate services. *Weather, Climate, and Society.*, **5** (**1**), 83–92.

38 Brugger, J. & Crimmins, M. (2012) *Weather, Climate, and Rural Arizona: Insights and Assessment Strategies*. Technical Input to the U.S. National Climate Assessment. CLIMAS, Tucson, AZ. http://www.climas.arizona.edu/publications/2586.

39 Brugger, J. & Crimmins, M. (2013) The art of adaptation: living with climate change in the Rural American Southwest. *Global Environmental Change*, **23** (**6**), 1830–1840.

40 Bidwell, D., Dietz, T. & Scavia, D. (2013) Fostering knowledge networks for climate adaptation. *Nature Climate Change*, **3**, 610–611.

CHAPTER 5

Participatory, dynamic models: a tool for dialogue

Laura Schmitt Olabisi[1], Stuart Blythe[2], Ralph Levine[1], Lorraine Cameron[3] and Michael Beaulac[4]

[1] Department of Community Sustainability, Michigan State University, 220 Trowbridge Rd, East Lansing, MI 48824, USA

[2] Writing, Rhetoric and American Cultures, Michigan State University, 220 Trowbridge Rd, East Lansing, MI 48824, USA

[3] Michigan Department of Community Health, Division of Environmental Health, 201 Townsend St., Lansing, MI 48913, USA

[4] Michigan Department of Environmental Quality, Executive Division, 525 West Allegan St., Lansing, MI 48909-7973, USA

5.1 Introduction: participatory, dynamic models as boundary objects

In supporting decision-making, Regional Integrated Sciences and Assessments (RISAs) serve as mediators between different professional communities (e.g., scientists and decision-makers), a role often referred to as a "boundary organization" [1]. RISAs are challenged to support decision-making that is relevant to the dynamics of climate change, human health, and infrastructure at local and regional levels. Acting as an intermediary in this setting often involves developing tools and products that represent the ways in which the components of a human–natural system interact, and that foster dialogue among scientists, decision-makers, and affected populations. Tools and products with these characteristics are called "boundary objects" [1].

A boundary object is a sociological construct that is often applied in interdisciplinary environmental research [2]. Boundary objects provide a translational function across disciplinary, cultural, or organizational divisions. To be effective, they must represent a problem in a manner that not only retains its essential structure across these divisions, but also

Climate in Context: Science and Society Partnering for Adaptation, First Edition.
Edited by Adam S. Parris, Gregg M. Garfin, Kirstin Dow, Ryan Meyer, and Sarah L. Close.
© 2016 John Wiley & Sons, Ltd. Published 2016 by John Wiley & Sons, Ltd.

allows the problem to be interpreted and manipulated by different users in different ways [3]. Many types of boundary objects have been used in environmental research, from concepts, to maps, to diagrams and decision calendars [Chapters 2, 1, 4, 5]. Similarly, a simulation model may be used and interpreted in different ways by different groups of people (e.g., a scientist may use the model for hypothesis testing, while a decision-maker may use the model for policy testing), but the underlying model structure, once negotiated and agreed upon, is not altered. We therefore propose that simulation models constructed using participatory methods can fit the role of a boundary object. In this chapter, we explain how Great Lakes Integrated Sciences & Assessments (GLISA) developed such a boundary object to assist Michigan decision-makers in planning for extreme heat events.

Participatory modeling has been practiced for several decades in multiple modeling traditions, but most extensively in system dynamics [6]. Although originally applied primarily to business and organizational management, participatory dynamic modeling is now becoming popular as a tool for navigating "messy" environmental problems [7,8]. A system dynamics model constructed in a participatory manner may become more than a decision-support tool; it may provide a framework for facilitating data collection and the exchange of information within important aspects of a complex system. When scientists and stakeholders build a system model together, they have the opportunity to learn together about the phenomenon being modeled, and also about the different assumptions that others bring to the task [9]. A model can also allow decision-makers to test proposed policies or actions in a simulation environment, and to examine the effectiveness of these proposals as well as the validity of the scientific reasoning behind them [10].

System dynamics modeling has several advantages that make it a suitable modeling approach for problems related to the environmental and human health impacts of climate change. It is a highly flexible modeling framework that can represent feedbacks and can integrate several types of data, including both qualitative and quantitative data [11]. System dynamics models are made up of stock-and-flow structures, so they are suitable for portraying problems such as carbon dioxide build-up in the atmosphere, or flows of people from a vulnerable to an infected population. Several software packages that are widely available for system dynamics allow the user to construct models using symbolic representations of stocks, flows, and other variables. These assist with participatory modeling, because users do not need to know how to program or read computer code in order to interpret or edit the model's structure. In addition, these packages come with the capability to develop user interfaces that allow for easy manipulation of key variables in the model. In the case study described below, we used Stella® software for the modeling work (iseesystems.com).

Typically, a participatory system dynamics model-building process begins with the group construction of a causal loop diagram (CLD) [6]. A CLD portrays the causal relationship between variables, including feedback loops. Construction of the diagram can itself be a highly valuable process, which produces key insights. Some participatory processes even consider the CLD to be an end product [6]). The CLD can, therefore, function as a boundary object, as participants in the diagramming process discuss the varied meaning each attaches to the elements of the diagram, and the nature of the causal relations depicted in the diagram [3].

However, if projections of future states of the system are a desired output from the modeling process, it is typically necessary to use the CLD as the template for a dynamic simulation model [6]. In this case, the model itself becomes the focal boundary object, as participants discuss its structure and behavior from their varied perspectives [12]. One of the most important functions such a model performs is that of making the assumptions of the various participants explicit, and testing the implications of these assumptions quantitatively. In constructing the CLD and the model, participants must effectively describe, using symbols and (sometimes) numbers, how they believe the system under study works. This representation of the system then becomes available to the group for discussion, debate, and critique. In traditional decision-making processes, these "mental models" of a given system often remain hidden, and conflict or disagreement arising from differences in these mental models is obscured, making consensus or enhanced understanding difficult to achieve [8].

Boundary objects often assist with core RISA functions, including the coproduction of knowledge between scientists, managers, and decision-makers; the translation of science into policy; mediation of contentious issues or situations; and convening stakeholders for a common purpose [13]. This type of facilitation and translation of knowledge to policy is a key component of the knowledge-to-action networks supported by RISAs (see Chapter 4 and Chapter 6). In this case study, we discuss how we applied participatory system dynamics modeling to the problems caused by impacts of extreme heat on human health in urban regions of the upper Midwest. Our project contains useful lessons on how similar efforts might be undertaken at RISAs in other parts of the country.

5.2 Modeling the human health impacts of extreme heat in Michigan

5.2.1 Problem background

Global and regional climate change models have projected an increase in the frequency, duration, and severity of extreme heat events in the upper Midwest. The frequency of hot days and the length of the heat-wave season will be more than twice as great under a higher emission scenario compared to a lower one [14]. This means that, by the end of the century (2070–2099) heat waves equivalent to the one that killed over 700 people in Chicago in 1995 are projected to occur about once every 3 years in the Midwest under the lower emissions scenario, and nearly three times a year under the higher emissions scenario [15].

It has been known for some time that extreme heat events can cause increased incidence of human illness, hospitalization, and death. The old, the young, and those with chronic health conditions are less able to perceive the stress on their bodies and are physiologically less able to withstand this stress [16]. In addition, social factors such as poverty and isolation may limit the possibility of some individuals to take actions to reduce their heat stress by, for example, moving to an air-conditioned building or shelter [17]. Extreme heat in the upper Midwest is therefore a deadly and worsening problem that affects already marginalized populations.

The impact of extreme heat on human health is a priority for many Michigan communities and for the Michigan Department of Community Health (MDCH). The Center for Disease Control's (CDC) Climate-Ready States and Cities Initiative funds the MDCH to develop ways to anticipate climate-related health effects by applying climate science, predicting health impacts, and preparing flexible programs. CDC encourages states and cities to collaborate with local and national climate scientists to understand the potential climate changes in their areas and incorporate knowledge into state and local decision-making related to adaptation to climate variability and change. The National Weather Service (NWS) predicts extreme heat events and issues appropriate warnings through local media. While these federal- and state-level entities offer support and scientific expertise, it is at the city or county health department level where most decisions pertaining to the health of citizens during extreme heat events are made. County and local health officials are in charge of making decisions about whether to open cooling centers (and how many, and in which locations); how to coordinate emergency care among local hospitals and clinics; and, how best to educate vulnerable citizens about the risks of extreme heat.

To explore the potential for supporting decision-making related to human health impacts of extreme heat in Michigan, GLISA supported a one-year scoping study using participatory system dynamics modeling. The team consisted of researchers from Michigan State University, the University of Michigan, the Michigan Department of Community Health (MDCH), and the Michigan Department of Environmental Quality (MDEQ). We were able to draw on ongoing data collection and outreach projects funded through the Centers for Disease Control's Climate-Ready States and Cities Initiative, administered through the MDCH.

5.2.2 Research questions driving the model

The questions we wished to explore through model construction were: How would an extreme heat event affect hospitalizations and mortality among elderly people in Detroit? How effective are various proposed policy or management strategies to reduce hospitalizations or mortality during a heat event? We elaborated on these questions in interviews with key stakeholders, as described below and seen in Table 5.1. There were three main phases of model development and evaluation.

5.2.3 Interviews for model construction

Because of the abbreviated timeline for the project (1 year) and limited budget, we interviewed key stakeholders to identify the variables and causal relationships that would define the model structure. Using interviews is accepted practice in participatory modeling, though not ideal, because it does not allow for significant interaction between stakeholders during the initial stages of model building [6]. However, we found it impractical to convene a model-building workshop early in the year without first gaining an understanding of the issues and stakeholders.

Eight key stakeholders with in-depth knowledge of heat events and their health impacts were selected from local contacts known to the project team, and included scientists from MDCH and MDEQ, academic experts, and county and city health officials. Interviews were semistructured, and focused on the interviewee's perception of the causal relations behind human health and extreme heat. Other topics covered in the interviews included potential policy responses to these heat events, the effectiveness of these policy responses, preventive measures that could be taken to reduce the human health impacts of extreme heat, and identification of vulnerable populations (see Table 5.1 for the interview guide). The interviews were recorded and transcribed, and the interview transcripts were analyzed for variables and interactions to be entered into the model.

Table 5.1 Interview questions used in the model-building stage.

---

1 Please tell us about the ways an extreme heat event can kill people, preferably in a chrono-logical way. Could you draw us a graph depicting deaths and hospitalizations per day, over time (Interviewer may provide an example to prompt)?

   a. Follow-up prompts: what climatic factors typically precede these deaths (tempera-ture/humidity/duration/etc.)?

   b. By what mechanisms do extreme heat events kill people?

   c. Over what period of time do we see mortality from these events?

   d. What makes a heat event "better" or "worse" in terms of the number of people who die?

2 Which groups of people are the most likely to die during an extreme heat event? Why?

3 When an extreme heat event happens, how do those affected typically behave/respond to becoming ill?

   a. How do hospitals or clinics behave/respond?

   b. How do local health departments behave/respond?

   c. How do emergency workers behave/respond?

   d. Are there any other local entities/groups that typically respond to heat events?

4 What strategies do you know of that have been used to prevent deaths from heat events before the event occurs?

   a. What resources do these strategies require? Who is responsible for implementing them?

   b. Have these strategies been effective? Why or why not?

5 What strategies do you know of that have been used to alleviate the severity of heat event mortality while the event is happening?

   a. What resources do these strategies require? Who is responsible for implementing them?

   b. Have these strategies been effective? Why or why not?

6 Some cities are developing warning systems for extreme heat events. Do you think these systems are likely to be effective? Why or why not? How far in advance of a heat event should a warning be issued?

7 Can you think of other strategies that might be effective at either preventing deaths or alleviating the severity of a heat event, but have not yet been tried? Why do you think they have not yet been tried? Why do you think they would be effective?

8 Finally, what other indirect human health impacts might extreme heat events cause? Over what period of time might we see these impacts?

Following the interview process, the two modelers on the project team assembled the interview notes and transcripts, and developed the causal structure of the model, which we named the Mid-Michigan Heat Model (MMHM), using Stella® software (Figure 5.1). Where interviewees were uncertain about, or disagreed on, a causal relationship (e.g., whether people would be more willing to use cooling centers if they could bring their pets with them), the modelers consulted the literature for any prior research that might give clarity to the topic. If no prior studies were found, the nature of

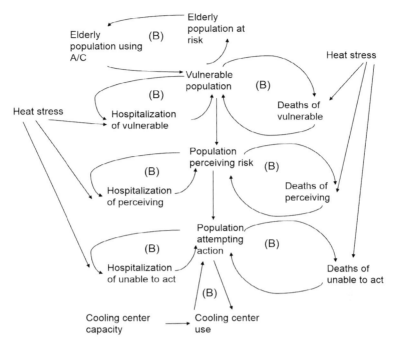

Figure 5.1 The causal loop diagram depicting the main structure and feedback loops in the Mid-Michigan Heat Model (MMHM). A series of balancing feedback loops regulate the flow of elderly people from being at risk, to being vulnerable (lacking air-conditioning), to perceiving the risk, and attempting action.

the causal relationship was considered to be one that could be tested with the dynamic simulation model. The model structure was presented to the full project team for feedback, two members of which had been interviewed during the model development phase. Edits and clarifications to the model structure were suggested and incorporated.

5.2.4 Model parameterization

After the project team agreed on the model structure, the modelers began collecting data to parameterize the model for a quantitative simulation run. Key data used to parameterize the model included a set of surveys on behavior during extreme heat in Washtenaw and Ingham counties, which targeted attendees of outdoor events in summer 2011. The modeling team also obtained heat event records from NWS for years 2001–2010, and hospitalization data for the same time period from Ingham, Washtenaw and Wayne counties, and metropolitan Detroit. The modelers attempted to use

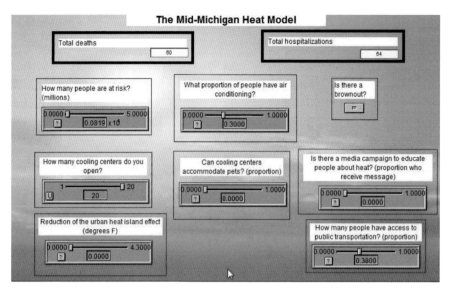

Figure 5.2 The Mid-Michigan Heat Model interface, with baseline settings indicated.

this data to derive a relationship between heat and hospitalizations during heat events in Michigan. However, the number of excess heat-related hospitalizations during periods of extreme heat in the Detroit area were very low, and did not exceed the model's margin of error. The modelers chose instead to use data from the 1995 Chicago heat wave episode to parameterize the causal relationship between heat and hospitalizations/deaths in the simulation model, because this heat wave had much more extreme health impacts than heat events in Michigan during the period 2001–2010 [17,18]. In order to check model output for internally consistent model behavior, the modeling team ran the semifinal version of the simulation model through several heat wave scenarios. Members of the project team also created a model interface to facilitate user interaction with the model (Figure 5.2).

5.2.5 Stakeholder workshop

The final, and most extensive, interaction with stakeholders took place during a daylong model evaluation workshop convened in May 2012. Workshop participants were suggested by the interviewees and the project team members, and all were members of organizations working on the problem of human health impacts of extreme heat (Table 5.2). Working with a facilitator, the project team introduced the MMHM and briefly described the system dynamics approach and modeling language. Workshop participants

Table 5.2 List of organizations represented at stakeholder workshop held in Detroit May 2, 2012.

Organization	Number of participants
Michigan State University	3
Great Lakes Integrated Sciences & Assessments Center	2
University of Michigan School of Public Health	5
University of Michigan Inter-University Consortium for Political and Social Research	1
Michigan Department of Community Health	5
Oakland County Health Division	1
Michigan Environmental Council	1
Washtenaw County Public Health Department	1
Detroit Office of Homeland Security and Emergency Management	2
National Weather Service Forecast Office, Detroit/Pontiac	1
Data Driven Detroit	1
Biomedware	2
Detroit Department of Health and Wellness Promotion	1
Detroit Office of Public Health Emergency Preparedness	1
City of Grand Rapids Planning Department	1
Michigan Department of Environmental Quality	1

were asked to discuss the health problems related to extreme heat, and their potential solutions, prior to interacting with the model. They then had the opportunity to interact with the model in small groups at their tables, with an experienced modeler facilitating at each table. Participants devised scenarios to serve as input for the model through the interface (Figure 5.2), and observed the output, which consisted of numbers of hospitalizations and deaths in urban Detroit during a heat wave. The model-building process described above was preliminary, and was intended only for scoping the problem of extreme heat and human health impacts. We therefore cautioned model users against interpreting the quantitative output of the model as anything other than an indicator of how the system behaves in response to input changes. Afterwards, they offered individual written feedback on the content and structure of the MMHM and on the workshop itself. Finally, a large group discussion focused on the utility of the MMHM and how it might be used in decision-making.

Feedback from the workshop was used to make significant revisions to the MMHM, including a reorganization of the interface. The modeling team also updated the model logic describing heat "watches" and "warnings," and created a more sophisticated structure describing vulnerable populations' willingness to take action during a heat event, based on participant feedback.

5.3 Lessons learned in the model-building process and their application to science in decision-making

Although being preliminary, the model yielded important insights. In the interviews and in the discussions that took place in the opening session of the workshop, most participants claimed that more cooling centers (air-conditioned, free public spaces where urban residents can shelter during heat events) would be the best way to fight heat-related illness and death. Country and city health officials, who were in a position to observe the use and effectiveness of cooling centers, felt universally that they were under-utilized. Interviewees posited different reasons for this under-utilization, from elderly urban residents' reluctance to leave their pets at home, to residents' fear of going to a cooling center located in a "bad" neighborhood, to lack of transportation.

All of these theories were tested on the model, but the most significant driver of cooling center use was vulnerable urban residents' perception of being in danger from extreme heat. Without this perception, vulnerable people lack the motivation to seek out options for combating heat illness, including going to a cooling center [19]. The model represents a behavioral pathway that is based on relevant social cognitive theory as applied in the health sciences (Figure 5.3). According to this theory, in order to change behavior (e.g., in response to a health threat), a person must consider an action, believe him or herself to be able to carry out this action, and have the means to do so [20]. These criteria are met in our model if a vulnerable person: (1) hears a message about a heat health risk that they believe is personally relevant; (2) is made aware of what to do about this risk; and (3) is able to act on these two messages.

This behavioral pathway was apparently more sophisticated than the mental models held by many workshop participants, who assumed that current levels or means of messaging are sufficient to bring vulnerable people to cooling centers (when, in fact, these may not be adequate to convince elderly people that heat health risks apply to them). Workshop participants therefore expressed surprise when they used the model settings to open more cooling centers, which then had no effect on overall deaths and hospitalizations. Although it is intuitive that, in order to act, urban

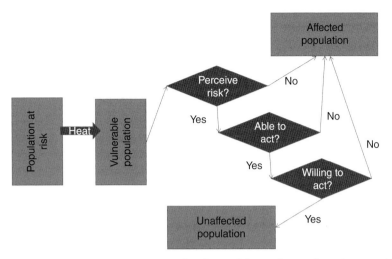

Figure 5.3 The decision structure represented in the model. In order to take action to avoid the consequences of extreme heat, the vulnerable population (elderly people in Detroit) must perceive a risk to themselves, be willing to act on that risk, and be able to act on that risk (e.g., by going to a cooling center). If one of these conditions is not met, hospitalization or death may occur.

residents must perceive a health threat and understand their options to mitigate that health threat, the model allowed workshop participants to experiment with the urban system and find out this truth on their own (Figure 5.3). In this respect, the model demonstrated an ability to call a widely held assumption into question (that opening more cooling centers automatically reduces the number of deaths and hospitalizations attributed to extreme heat).

The model also demonstrated extraordinary sensitivity to electrical brownouts. These had the effect of multiplying deaths and hospitalizations by a factor of two to three in the model output, through the mechanism of eliminating vulnerable peoples' ability to use air-conditioning. Interestingly, this highlights a critical negative feedback loop that affects urban residents' health during a heat event. As temperatures rise, more businesses and residences use their air-conditioning, which strains the electrical grid and can lead to a brownout, thereby threatening the health of vulnerable populations. This is an example of how a system dynamics model can expose the systemic nature of a problem, and its exogenous as well as endogenous drivers.

Workshop participants praised the model for its complexity, realism, and for the way it portrayed interactions between variables. They also appreciated that the model could portray the consequences of choices such as reducing the urban heat island effect, or opening more cooling centers, on the variables of interest (deaths and hospitalizations). They found it easy to interact with the model.

On the other hand, many workshop participants felt that they did not fully understand the assumptions behind the model, and the data and information that were used to build it. This was one of the drawbacks of using an abbreviated participatory process involving interviews with a select few stakeholders to build the model. When participants are involved in each stage of model development, they are more likely to understand the assumptions behind the model, and trust its output [21]. This lack of transparency could have been ameliorated somewhat by giving each participant a guide describing the model construction and data sources. However, this approach would have risked alienating some stakeholders by overloading them with technical information. The participatory modeling literature contains many case studies describing the attempt to achieve a balance between providing stakeholders with "too much" versus "too little" information about the technical aspects of the model, and the resolution of this problem is necessarily context-specific [22,23]. However, we believe there is cause for further research on how nonscientists (or scientists who are not modelers) perceive technical information about a model, as well as on optimal ways to communicate this information, as discussed in subsequent text.

Workshop participants also wanted the model to portray more spatial information, such as which parts of an urban area might be more susceptible to heat risk than other areas, or the specific neighborhoods where vulnerable populations live. Integrating spatial information into the model was not possible using the Stella® platform, but this represents a future key direction for the heat and human health modeling work. Other studies involving a participatory approach to spatial planning and modeling have found this to be an effective way to address highly complex and contentious problems compared with traditional planning approaches [24].

To summarize our experience with this prototype model, we found it very useful for challenging widely held, yet unexamined, assumptions about solutions to a systemic problem (cooling centers as an intervention for preventing deaths and hospitalizations from extreme heat). The model was also helpful for illuminating the systemic nature of human health impacts from extreme heat, and their dependence on energy systems (the electrical grid) and human behavior (vulnerable populations' perceptions of risk and decisions to act). Both of these types of insights are consistent with previously reported benefits of using participatory system dynamics modeling in an environmental context [8,10,25]. MMHM acted as a boundary object in revealing these aspects of system behavior to the workshop participants and to the scientists involved in building the model. Each participant was able to understand this behavior and interpret it in the context of their own work, as exemplified by the participant comments and discussion at the end of the workshop. If we had used a mediated discussion approach to addressing problems related to heat and human health, the conversation would likely not have gone very far past how many cooling centers should be opened and where to place them. This is because most stakeholders appeared not to consider the social cognitive motivations behind vulnerable peoples' decision to take action to protect themselves during extreme heat [20]. The model opened up the conversation about heat and human health to include a much wider range of environmental and behavioral stressors than might otherwise have been considered by the decision-makers at the workshop.

Based on our experience, we believe that participatory dynamic modeling represents a promising tool for building stakeholder consensus and revealing the dynamic and systemic nature of environmental problems. However, there are some key constraints to the application of this type of modeling to the climate change science and decision-making supported by RISAs and other boundary organizations. The incorporation of spatial information into the modeling framework represents one such constraint. Many climate variability and change-related environmental problems have critical spatial dynamics associated with them. Most system dynamics software packages are ill-equipped to handle spatial data, making it difficult to represent these dynamics using a participatory modeling approach. Dynamic modeling can be done with most GIS software packages but, typically, the models built in these software environments are opaque to stakeholders who do not read or understand computer code. There is a need for more work on designing and implementing participatory spatial modeling processes.

Second, logistical constraints to a participatory modeling approach must be considered. Building a dynamic simulation model with the full participation of relevant stakeholders is time consuming and costly. We were able to complete a participatory modeling process in only 1 year and with a modest budget, but we did not build the model structure with the full group of stakeholders who represented the clientele for our model. This would have taken at least two to three workshops to accomplish [6]. As mentioned above, even a CLD can function as a boundary object, which can generate systemic insights about a problem, although it lacks quantitative rigor and the ability to demonstrate how dynamics unfold over time. A CLD may be constructed in a half-day workshop, making it appropriate for projects with a shorter time span and a smaller budget.

A third challenge to the widespread use of participatory modeling techniques is the lack of accepted practice for conducting and evaluating these processes. Participatory dynamic modeling is a relatively new methodology in the research for environmental problems. Protocols for conducting participatory modeling exercises and assessing their effectiveness at providing the translational functions of a boundary object have been developed mostly as isolated case studies [12]. This is a problem, because it represents a barrier to learning about the process for those who might wish to use participatory modeling methods, but who have no experience doing so. Developing protocols for participatory modeling efforts related to adaptation to climate change could allow for more cross-regional and cross-study comparisons of these efforts.

5.4   Lessons for institutional design of boundary organizations

The case study described above contains valuable lessons on how participatory modeling may be best used as one tool in the RISA "toolbox" for enhancing the use of climate science in decision-making. Based on our experience with this project, and on the growing literature on participatory modeling of environmental problems, we believe that model building that takes place jointly between scientists and stakeholders can generate a model that could become a boundary object. Its characteristics include fostering a common understanding of complex environmental problems and providing a shared learning experience [9]. Participatory model building can also help to focus future scientific inquiry—as in this case study, in which we learned that the dynamic spatial patterns of heat vulnerability are important to decision-makers. In contrast with other consensus-based decision-making processes, participatory modeling can help to challenge decision-makers' commonly held beliefs that may not be accurate or may not reflect a complete understanding of the system [8]. In the MMHM project, this was demonstrated through the model's challenge to the notion that more cooling centers would necessarily prevent heat illness and deaths.

In order to use participatory model building in an effective and replicable way in boundary organizations or knowledge-to-action networks it would be beneficial to recruit scientists or practitioners with participatory model-building skills as P.I.'s or researchers. As mentioned above, establishing a set of protocols or best practices for participatory modeling could help further both the application of climate science to decision-making and the field of participatory modeling. Because it is time-consuming and expensive, participatory model building should be used strategically and judiciously. It is most useful for addressing "wicked" or "messy" problems that involve not only complex science, but also multiple stakeholder perspectives and disagreements on values [7]. Climate variability and change science and adaptation, of course, provide many examples of such problems. For a situation in which building a full dynamic simulation model may be infeasible or inappropriate, a CLD is easier to construct and may also function as a boundary object that can provide systemic insights.

Finally, an area of participatory model building that has been under-researched is the design of materials used to communicate information about models and the products they generate with model-building participants and with the broader public. In this project we found that, although most workshop participants commented positively about the model interface, much more thought could have been devoted to its design and the ease of its use. For example, it was not immediately clear to model users which functions on the interface should be operated first, or how one should go about constructing a model scenario to test. In addition, several workshop participants expressed the desire for other materials, such as a manual, that could accompany the model and allow them to better understand it. We suggest the inclusion of experts in scientific and technical communication, who can help to design such materials, in any large model-building process.

By involving decision-makers in the model-building process, we hoped to change the way they thought about the problem of heat and human health in urban Detroit, leading to systemic insights that would help them design more effective interventions. Feedback from the workshop suggests that many participants did learn from their interactions with the model. The limited scope of this project did not allow us to track long-term shifts in participants' thinking or decision-making, although other studies have demonstrated that learning from participatory model building is real and persistent over time [26]. This modeling process was not linked with a pressing policy decision, which may have allowed us to observe the impacts of participatory modeling on actions more directly. However, participatory modeling may not be able to overcome the entrenched ideological attitudes that stakeholders often develop in the face of a pressing and controversial decision process [23]. Therefore, using an exploratory model early in the knowledge-to-action process may be more effective at informing decisions than using such models closer in time to a decision point—although the pathway from knowledge to action may be more diffuse.

5.5 Conclusions

Joint model building between scientists and stakeholders has the potential to transform knowledge to action through orienting model building around questions stakeholders care about; enhancing stakeholders' understanding of system properties such as feedback, non-linear dynamics, and uncertainty; explicitly stating and debating assumptions underlying policy and management decisions; and simulating policy or management decisions under different scenarios to observe their effectiveness. In order to use these tools effectively, knowledge-to-action networks should recruit or train modelers with experience in participatory model-building techniques, and support participatory model-building efforts with adequate funding and personnel. Modeling may be used early in the decision process for decision scoping, which helps to ensure that policy-makers and scientists are asking the right questions. Finally, supporting materials and the model interface should be designed for maximum legibility and clarity.

References

1 Guston, D. (2001) Boundary organizations in environmental policy and science. *Science, Technology, & Human Values*, **26** (**4**), 339–408.

2 Star, S.L. & Griesemer, J.R. (1989) Institutional ecology, 'translations' and boundary objects: amateurs and professionals in Berkeley's Museum of Vertebrate Zoology, 1907-39. *Social Studies of Science*, **19**, 387–420.

3 Black, L.J. & Andersen, D.F. (2012) Using visual representations as boundary objects to resolve conflict in collaborative model-building approaches. *Systems Research and Behavioral Science*, **29**, 194–208.

4 Harvey, F. & Chrisman, M. (1998) Boundary objects and the social construction of GIS technology. *Environment and Planning A*, **30**, 1683–1694.

5 Brand, F.S. & Jax, K. (2007) Focusing the meaning(s) of resilience: resilience as a descriptive concept and a boundary object. *Ecology and Society*, **12**, 23–39.

6 Vennix, J.A.M. (1996) *Group Model Building. Facilitating Team Learning Using System Dynamics*. John Wiley & Sons, New York.

7 Vennix, J.A.M. (1999) Group model-building: tackling messy problems. *System Dynamics Review*, **15** (**4**), 379–401.

8 Van den Belt, M. (2004) *Mediated Modeling: A System Dynamics Approach to Environmental Consensus Building*. Island Press, Washington, D.C..

9 Schmitt Olabisi, L., Kapuscinski, A.R., Johnson, K.A. *et al.* (2010) Using scenario visioning and participatory system dynamics modeling to investigate the future: lessons from Minnesota 2050. *Sustainability*, **2**, 2686–2706.

10 Costanza, R. & Ruth, M. (1998) Using dynamic modeling to scope environmental problems and build consensus. *Environmental Management*, **22**, 183–195.

11 Hirsch, G., Miller, R. & Levine, R. (2007) Methods for understanding the impact of social change policy initiatives: an illustration of the system dynamics modeling approach. *American Journal of Community Psychology*, **39**, 239–253.

12 Hovmand, P.S., Andersen, D.F., Rouwette, E. *et al.* (2012) Group model-building 'scripts' as a collaborative planning tool. *Systems Research and Behavioral Science*, **29**, 179–193.

13 Parris, A., Simpson, C. & Abdelrahim S. (2010) *RISA Workshop Report: Looking Ahead at Climate Service, Assessment, and Adaptation*. NOAA, Washington, DC.

14 Pryor, S.C., Scavia, D., Downer, C. *et al.* (2014) Chapter 18: Midwest. In: Melillo, J.M., Richmond, T.C. & Yohe, G.W. (eds), *Climate Change Impacts in the United States: The Third National Climate Assessment*. U.S. Global Change Research Program, Washington, DC, pp. 418–440. DOI:10.7930/J0J1012N.

15 Hayhoe, K., Sheridan, S., Kalkstein, L. *et al.* (2010) Climate change, heat waves, and mortality projections for Chicago. *Journal of Great Lakes Research*, **36 (Supplement 2)**, 65–73.

16 Khosla, R. & Guntupalli, K.K. (1999) Heat-related illnesses. *Critical Care Clinics*, **15**, 251–263.

17 McGeehin, M.A. & Mirabelli, M. (2001) The potential impacts of climate variability and change on temperature-related morbidity and mortality in the United States. *Environmental Health Perspectives*, **109**, 185–189.

18 Semenza, J.C., McCullough, J.E., Flanders, W.D. *et al.* (1999) Excess hospital admissions during the July 1995 heat wave in Chicago. *American Journal of Preventive Medicine*, **16**, 269–277.

19 Abrahamson, V.J., Wolf, I., Lorenzoni, B. *et al.* (2009) Perceptions of heatwave risks to health: interview-based study of older people in London and Norwich, UK. *Journal of Public Health*, **31**, 119–126.

20 Bandura, A. (2004) Health promotion by social cognitive means. *Health Education & Behavior*, **31**, 143–164.

21 Stave, K.A. (2002) Using system dynamics to improve public participation in environmental decisions. *System Dynamics Review*, **18**, 139–167.

22 Prell, C., Hubacek, K., Reed, M. *et al.* (2007) If you have a hammer everything looks like a nail: traditional versus participatory model building. *Interdisciplinary Science Reviews*, **32 (3)**, 263–282.

23 Zellner, M.L., Lyons, L.B., Hoch, C.J. *et al.* (2012) Modeling, learning, and planning together: an application of participatory agent-based modeling to environmental planning. *URISA Journal*, **24**, 77–92.

24 Jordan, N.R. *et al.* (2011) TMDL planning in agricultural landscapes: a communicative and systemic approach. *Environmental Management*, **48 (1)**, 1–12.

25 Hare, M., Letcher, R. & Jakeman, A. (2003) Participatory modelling in natural resource management: a comparison of four case studies. *Integrated Assessment*, **4**, 62–72.

26 Scott, R.J. *et al.* (2013) *Evaluating Long-Term Impact of Qualitative System Dynamics Workshops on Participant Mental Models*. 31st International Conference of the System Dynamics Society. Cambridge, MA.

CHAPTER 6

Not another webinar! Regional webinars as a platform for climate knowledge-to-action networking in Alaska

Sarah F. Trainor[1], Nathan P. Kettle[2] and J. Brook Gamble[1]

[1] *Alaska Center for Climate Assessment and Policy, University of Alaska Fairbanks, 505 S Chandlar Drive, Fairbanks, AK 99775, USA*

[2] *Alaska Center for Climate Assessment and Policy, University of Alaska Fairbanks, and Alaska Climate Science Center, 505 S Chandlar Drive, Fairbanks, AK 99775, USA*

6.1 Introduction

Webinars, or web-based seminars, have rapidly become a popular format for disseminating information, hosting professional development programs, and promoting collaboration. In the climate realm alone, the number of webinars offered has increased dramatically in recent years,[1] expanding the topics and amount of information available, as well as contributing to potential "information overload." Digital information technologies have been shown to contribute to the creation of virtual communities and bridging collaboration across geographic space and time zones. They also enhance networks by increasing trust and group solidarity. Despite the increasing use of webinars, there remains a limited understanding of the role of long-term webinar series in supporting knowledge-to-action networks (KANs) for climate change adaptation. For the purposes of this chapter, we use the term KAN to refer to an array of individuals and organizations operating at multiple scales, whose collaborative networking connects climate science, services, and applications for climate adaptation [1].

[1] The increase in climate webinars was detected based on the analysis of data obtained from an extensive web-based search using a Google search with combinations of the following keywords: climate, webinar, web-conferencing, series, and climate change.

Climate in Context: Science and Society Partnering for Adaptation, First Edition.
Edited by Adam S. Parris, Gregg M. Garfin, Kirstin Dow, Ryan Meyer, and Sarah L. Close.
© 2016 John Wiley & Sons, Ltd. Published 2016 by John Wiley & Sons, Ltd.

Established in 2006, the Alaska Center for Climate Assessment and Policy (ACCAP) Regional Integrated Sciences and Assessments (RISA) collaborates with stakeholders to inform realistic community plans and climate adaptation strategies using the most scientifically accurate, reliable, and up-to-date information. ACCAP initiated the Alaska Climate Webinar Series[2] in June 2007 with the goals to: (1) create a forum for discussion of specific aspects of climate change in Alaska, (2) identify information gaps and how best to fill them, and (3) improve the ability of Alaskans to prepare for and respond to climate change. Due to the relatively rapid rate of warming in Alaska and the plethora of related impacts on both natural and human systems [2], this webinar series offers information and decision-support tools, which address a broad diversity of climate-related topics, sectors, specific stakeholder interests, and sub-regions within Alaska.[3] We begin this chapter with a discussion of the importance of knowledge-to-action networks in supporting adaptation and a description of the ACCAP Climate Webinar Series. We engage in a combination of interviews and participant observation to document the significance of the ACCAP webinar series in fostering the evolution of a knowledge-to-action network, both topically in the subjects and information presented and procedurally in the development of a network of participants and speakers. The chapter also includes reflections on the process of using webinars as a tool for creating and building knowledge-to-action networks.

6.1.1 Knowledge-to-action networks

Social networks provide several functions that support climate adaptation. On a foundation level, networks serve as a mechanism for the exchange of data, information, and knowledge among participants [3,4]. Networks also provide opportunities to build collaborative partnerships, which enable members to pool resources, share lessons learned and best practices, and facilitate ongoing relationships within and across organizations and agencies [5]. Bringing together multiple groups may enhance credibility, legitimacy, and saliency, though there are often trade-offs that must be mediated [6]. Additional functions include distributing risks, building consensus, identifying common goals, negotiating conflicts, and providing critical capacity for action, such as supporting public education and outreach or data collection, monitoring, and research [4,7]. These functions

---

[2] Also referred to in this chapter as the ACCAP Climate Webinar Series or the ACCAP webinar series.

[3] A complete listing of past and upcoming webinar topics and speakers is available from the ACCAP website at http://accap.uaf.edu/?q=webinars.

contribute to environmental and network governance across multiple levels of management [8].

Social networks can take several forms based on the density of their connections and level of formality [8]. Some networks have a core group of membership characterized by a high density of connections and stable network membership, while others are more diffuse. Regarding the level of formality, some networks are formal and based on regulatory frameworks and official partnerships, and others are more spontaneous, ad hoc, or self-organized. Both formal and informal shadow networks are important in responding to, and building the capacity to address, multilevel governance challenges such as climate change [9].

Supporting knowledge-to-action networks requires attention to site-specific decision-making contexts [10]. Within coupled human–environmental systems, decision-making contexts refer to a host of social, psychological, and physiographic factors that influence the decisions of individuals, organizations, or sectors in order to achieve their goals and objectives [11]. These contexts include nonclimate factors such as population and land use change, governance regimes, institutions, political and legal issues, and other barriers and limits. They also include the interaction of climate variability and change with existing management stressors. Decision-support initiatives designed to support knowledge-to-action networks are therefore more likely to reach long-term and sustaining strategies when they are applied to a wide range of specific decision-maker needs, which vary by sector, season, and geographic region [3,12], and when they engage a diversity of participants such as climate scientists, boundary organizations, and decision-makers (see Chapter 4). Such processes also contribute to the coproduction of usable knowledge and facilitate a continued two-way interaction between scientists and decision-makers [13]. Emerging research also suggests that fostering extended network connections among researchers and decision-makers is critical for supporting climate-sensitive decisions in order to keep up with the rapidly changing actors, connections, barriers, and information needs. Such networks between scientists and stakeholders are critical in generating usable science [14].

6.1.2 The ACCAP Alaska Climate Webinar Series

The ACCAP Climate Webinar Series was designed, organized, and created by ACCAP to support knowledge-to-action networks across Alaska. Between June 2007 and June 2013, ACCAP hosted 68 climate-related webinars. The webinar format in which participants can join remotely or as part of hub satellite sites located around the state promotes both virtual and face-to-face interactions. ACCAP publicizes the webinars via the ACCAP listserve and social media (Facebook and Twitter), and announcements submitted to other organizations such as the Arctic Institute, the Arctic Research Consortium of the US (ARCUS), the National Oceanic and Atmospheric Administration (NOAA), the University of Alaska Fairbanks (UAF) media relations, and various UAF departments for announcement in their print, email and on-line calendars, newsletters, and listservs. Satellite hub locations also publicize through their own networks. Webinars occur on a monthly basis and typically last 1 hour, which includes an introduction to the webinar series, the presentation by the invited speaker, and a question and answer session. A wide range of topics are included in the Climate Webinar Series, reflecting the diversity of climate-related challenges and opportunities across the state. Webinars showcase cutting edge scientific research results, innovative climate-related decision support tools, and national-scale reports and initiatives. Topics are selected to create a balance among marine/coastal, terrestrial, and human dimensions research, and cover themes related to adaptation and planning, climate variability and change, and climate impacts (Figure 6.1). Speakers are invited from a range of universities nation-wide and from organizations and agencies across the state and the country. Our target audience, representing our major stakeholder groups, is diverse and varied, including communities, tribal governments, municipal, state, and federal agencies, nongovernmental and nonprofit organizations (NGOs), the news media, scientists and engineers, planners and managers, and industry statewide in Alaska.

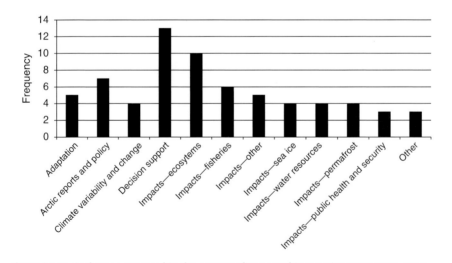

Figure 6.1 Topic themes presented in the ACCAP Climate Webinar Series (June 2007–June 2013). The label "Impacts" refers to climate impacts. The label "Other" includes presentations by representatives from the National Science Foundation highlighting funding opportunities relevant to climate and environmental change in the Arctic as well as a presentation by and about the US Fish and Wildlife Landscape Conservation Cooperatives in Alaska.

ACCAP also works to ensure that webinar topics are timely and relevant. For example, late summer webinars might focus on the latest research relevant to wildfire and sea ice extent. Webinars have also been hosted at the request of organizations such as the National Science Foundation, the U.S. Arctic Research Commission, and the U.S. Global Change Research Program (USGCRP) to solicit stakeholder input on plans, reports, and documents and to link scientists with stakeholders.

Several steps are taken to enhance the accessibility of webinars through the consideration of existing telecommunication infrastructure, software, and presentation archiving. For example, webinars are set up with separate audio and visual components, which is germane for rural villages in Alaska, that have limited internet bandwidth and connectivity. Webinar slides are available for download on the ACCAP website so that participants can view the slides in the event they are not able to stream the presentation over the internet. Participants are instructed to direct their internet browser to a URL, rather than requiring the installation of software on their personal computers, which is especially problematic for government officials who may not have administrative permissions to make changes on state-owned computers. Finally, presentations are recorded and archived on the ACCAP website to increase the accessibility of the presentation for those interested in attending, but are not able to participate in real-time.

Participants attend webinars individually or with a group at a "satellite hub site," where multiple participants gather to attend. Satellite sites were initially organized in the winter of 2009 as a way for ACCAP to extend the reach of the webinar series to additional audiences and promote partnering relationships with organizations statewide. Currently, there are seven active hub sites around the state that participate, and participation ranges from two to four hub sites per webinar. In these locations, a host contact is identified and ACCAP provides additional publicity material. Additional satellite sites are self-organized by interested and motivated participants.

6.1.3 Understanding the context

The State of Alaska spans approximately 580,000 square miles, including 44,500 miles of coastline. This landscape includes diverse ecoregions such as coastal rainforests, intermontane boreal, and arctic tundra. The majority of the state's population (700,000) lives in the major population centers of Anchorage and the surrounding vicinity, Fairbanks, and Juneau. The remaining population is dispersed among over 300 communities, most of which have primarily indigenous residents and are accessible only by river or air. Major economic sectors across the state include oil and gas, minerals, seafood and fishing, and tourism. The diversity of ecoregions, economic sectors of importance, and dispersed population highlight the wide range of management challenges throughout Alaska.

Over the past century, there have been notable climate-related changes across Alaska, including changes in temperature, precipitation, and sea ice extent [2]. Mean statewide temperatures have increased by 3 °F since 1949, with the greatest increase occurring during winter months.[4] Analysis of climate model outputs suggests that the rate of change for several climate variables is projected to accelerate across much of the state throughout the 21st century. For example, regional climate forecasts (CMIP 3, both A1 and B2 scenarios) suggest that mean annual temperatures may increase by 1.5–4.5 °F by 2021–2050, with geographic variability across the state and greater increases in the north. This is nearly twice the rate of many areas in the conterminous United States. These changes have contributed toward several environmental changes, such as increased coastal erosion, permafrost melt, sea ice loss, wildfires, and ocean acidification, which pose a potential of high risk to several economic sectors, government services, and subsistence food harvest across Alaska [2].

There is a wide array of actors working in areas related to multiple aspects of climate and climate change across Alaska, all with varying levels of experience and expertise with climate-related issues. These include land and resource managers and planners at multiple levels of governance, individuals and organizations involved in providing climate services and research, as well as citizens whose traditional ways of life and living off the land are changing. The growing interest in climate variability and change across Alaska and the United States has led to increasing support for and development of entities, organizations, and partnerships across Alaska [2,15]. Together, the above factors highlight the urgent need for supporting knowledge-to-action networks for climate-sensitive concerns in Alaska and the site-specific conditions within the socio-ecological system that must be considered in the design of decision-support activities.

6.2 Methods

Analysis of the ACCAP Climate Webinar Series is based on examination of two data sets: the webinar participant database (June 2007–June 2013) and semistructured interviews (2010 and 2013). Webinar participation data were

---

[4]Alaska   Climate   Research   Center: http://climate.gi.alaska.edu/ClimTrends/Change/TempChange.html [accessed on 9 August 2013].

analyzed to understand overall trends in participation. Early in the webinar series (2007–2009), participation data were collected by documenting names and affiliations as participants introduced themselves at the beginning of the webinar. With the introduction of new web-conferencing software, participant lists can be automatically generated and downloaded after the event (2010–present). Prior to analysis, data were cleaned by editing name and organization mis-spellings and coding participant organizations into different categories (college/university, NGOs, local government, state government, federal government, Alaska Native, media, satellite sites, other, and no data). Satellite sites are locations with three or more participants viewing a webinar from the same location.

Two sets of semistructured interviews were used to understand the role of ACCAP climate webinars in facilitating knowledge-to-action networks (2010 and 2013). There were 11 and 14 participants for the 2010 and 2013 interviews, respectively. Participants were selected based on a purposeful sampling technique, which ensured that participants represented a range of backgrounds, including organization type, frequency of attendance (single and multiple webinars attended), and geographic region of Alaska. Interviews typically lasted between 45 minutes and 1 hour and were digitally recorded. Audio files were transcribed and NVIVO content analysis software was used to code each of the transcripts for themes related to information dissemination and network processes and evaluation. Themes included both individual and network-related actions resulting from webinar participation, factors influencing development of knowledge-to-action networks (e.g., time/resource constraints, institutional change, development of general knowledge base, and provision of archives), and type of network link (e.g., participant to participant, participant to non-participant, and participant to speaker). Each transcript was reviewed by project staff to validate the transcription process.

6.3 Findings

We begin this section with a report of the findings regarding webinar participation and then outline the findings related to *outcomes*, or results that occur, in part, based on participation in ACCAP webinars. These outcomes include learning about the substantive content of climate science, impacts, and adaptation (*knowledge*), using climate information obtained in the webinar in decision-making and outreach (*action*), and fostering the development of networks to advance member success, promote dialogue, and help connect participants to decision-maker priorities, concerns, and needs (*network processes*).

6.3.1 Webinar participation

Participation in the ACCAP Climate Webinar Series is characterized by an expanding, fluid, and diverse network of participants. Participation per webinar has nearly tripled since the inception of the webinar series in June 2007. The mean attendance for 2013 was roughly 80 participants per event. Total attendance at ACCAP webinars between June 2007 and June 2013 exceeded 1400.[5] The majority of participants have attended one or two webinars (73% and 13%, respectively), suggesting a fluid network membership. Analysis of the interview data suggests that the high number of participants attending a single webinar occurs in part because of the breadth of topics covered in the webinar series, not because participants were dissatisfied with the experience.

Webinar participants have a wide range of backgrounds and experience with climate-related issues. The majority of participants are from federal agencies (40%); however, there is also notable participation from NGOs (8%) and colleges/universities (7%) (Figure 6.2). An analysis of participants' organizational affiliation revealed that individuals across the state of Alaska attended the webinar series. Some participants had worked with climate-related issues for well over a decade, while others were very new to the topic of climate change. Several interviewees commented on the diversity of participants attending ACCAP webinars. For example, one participant remarked,

> I'm just always really amazed at the diversity of the participants. There's everyone from people in villages further out than Kotzebue to people ... all the way out (sic) to Washington DC.
>
> *(Participant 1, USFWS, 2010)*

Although the majority of participants attended webinars via personal computers, the telephone, or both (74%), others participated in webinars at satellite hub sites. Satellite site attendance rates, hosting organizations, and types of participants at individual sites vary greatly. Attendance at satellite sites ranged from 3 to 45 individuals, although attendance was most commonly between 3 and 15 people. Several of the ACCAP-organized satellite sites are hosted by the federal government or federally funded agencies, such as the Alaska Pacific River Forecast Center (APRFC), the

---

[5] The total number of unique participants was calculated at 708, but the true number is likely to be higher because of missing data. For example, some participants did not provide their name or affiliation at login and were therefore coded as "no data" ($n = 638$) and excluded from the total count.

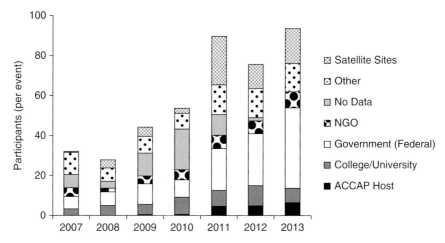

Figure 6.2 ACCAP webinar participation (per event). The "Other" category represents media, industry, Alaska Native organizations, local and state government, and international entities.

Alaska Ocean Observing System (AOOS), and Kachemak Bay National Estuarine Research Reserve (KBRR). Satellite sites are also hosted by the Sea Grant Marine Advisory Program specialists, Cooperative Extension agents, NGOs, and rural campuses in the University of Alaska system. Some satellite sites, such as the APRFC, host participants who are all individuals from the same organization or agency, while other sites, such as the KBRR, publicize widely throughout the local community and host a diversity of local participants.

6.3.2 Knowledge transfer

Interviewees reported that they found the webinar series to be a useful and informative venue for acquiring specific information about climate science and impacts, as well as learning about the availability and application of specific decision support tools. This includes information about changing environmental conditions and social systems as well as linkages between climate and non-climate-related resource management issues. Obtaining up-to-date information on climate science and decision-support specific to the Alaska region and communities is one of the primary motivating factors that influence ACCAP webinar attendance. One webinar participant, for example, stated that ACCAP webinars provided a unique opportunity to "gain new information and stay on top of current climate-related science within the state" (Participant 2, KBRR, 2013). As the following quotations illustrate, climate researchers, service providers, the media, and other key decision-makers all expressed similar views that ACCAP webinars are a valuable source for climate-related information:

> I think it's a fantastic opportunity to learn more about what the state of science is and what kind of capabilities we have, especially when we are faced with travel restrictions and budget restrictions.
>
> *(Participant 1, NGO, 2013)*

> If the topic sounds like it would be of interest to me and to a general newspaper audience, then I've found it's a good way to kind of get a quick dose of a scientific overview on that subject in a hurry.
>
> *(Participant 3, Media, 2013)*

In addition, several interviewees stated that the ACCAP webinar series was effective in increasing their knowledge of climate projections and impacts across Alaska. The March 2013 webinar on the Alaska Chapter of the National Climate Assessment (NCA), for example, was especially helpful in this regard. One participant, a newspaper journalist, stated:

> I learned a lot from that [webinar], you know you hear a lot about various scenarios for warming and it was interesting to get all of those encapsulated into one place because I feel like at times there are competing scientific visions for what's going to happen. So it was nice to get that under an umbrella of a larger report...I feel like it gives me a foundation or an overview for a specific subject.
>
> *(Participant 3, Media, 2013)*

Participants also gain information about specific climate impacts in Alaska that are associated with changing environmental conditions (e.g., permafrost thaw and ocean acidification) and social systems (e.g., infrastructure, food security, and public health), which help in operational management decisions. For example, a hydrologist for the APRFC reported that webinar participation deepened her understanding of the impact of permafrost thaw on soil infiltration. Such information is useful to her for inferring how changing hydrologic conditions may affect parameterizing hydrologic models and thereby forecast operations.

Webinars are also a source of information about links between climate, resource management issues, and other factors. For example, one webinar participant who has worked with climate-related issues for the past 10 years stated that,

> I don't know about forestry, but it [the ACCAP webinar] ended up mixing the climate and the forestry and it worked very well; and I learned a little bit about ... all these (sic) things climate affects directly, so I learn something other than climate while I'm listening to these.
>
> *(Participant 11, College/University, 2013)*

Since many webinars are focused on showcasing new decision-support tools, participants learn about new and relevant data portals and mapping applications for Alaska. Interviewees reported learning about where to find real-time information on weather, ocean conditions, and oil spills, which were available online from organizations such as the University of Alaska's Geographic Information Network for Alaska (GINA), AOOS, and NOAA's Environmental Response Management Application. Webinars are also effective in disseminating information about national-scale reports, projects, and funding opportunities to an Alaskan audience. For example, one participant in a rural village in Alaska commented that she would not have had the opportunity to hear about the findings from the recently released Arctic Integrated Management Report, if not for the ACCAP webinar showcasing report findings. Together, these examples illustrate how the webinar series promotes learning about climate science, impacts, and decision-support tools.

6.3.3   Action

Webinar participants reported several different types of actions that ensued from webinar participation. These include using climate information obtained from the webinar in trainings, education, planning, decision-making, and outreach. As outlined in Section 6.5, however, more work is needed to document and analyze actions that have resulted from webinar participation.

Interviewees reported using information acquired in ACCAP webinars directly in their job responsibilities by incorporating content materials from webinars into education and trainings, resource planning, and private consulting. For example, a Regional Education Coordinator for the National Park Service (NPS) discussed the importance of acquiring new figures, graphics, and other information that have been used in the development of training materials for park interpreters and for emerging public education and climate literacy efforts.

> [The ACCAP webinars series has] certainly been very supportive in terms of helping me develop training [materials]. I've developed web-based training tutorials for park interpreters and communication folks to learn about climate change ... In some cases I've actually used graphics, visuals, or slides from the presentations, through the speakers' permission, to actually incorporate into the training presentation ... most of our focus has been on using the materials to help train our frontline interpreters and frontline education specialists so they can deal informally with questions by visitors when they come up.
> *(Participant 6, NPS, 2013)*

In another example, information presented in a webinar was incorporated into a USFWS wildlife refuge conservation plan. A private sector consultant reported using information on ice break-up that was presented in a webinar to advise clients about the timing of construction and field work.

Finally, ACCAP webinars assist the news media in synthesizing scientific information for the public and identifying leading scientific experts as future contacts.

> Every ... [webinar] I have attended has at least resulted in an article ... that provides a summary of the science that was discussed in the webinar. I think in several of them I have also become familiar with scientists that I have used as contacts after the fact on a particular subject.
>
> *(Participant 3, Media, 2013)*

While more research is needed to document and analyze direct action outcomes from webinar participation, this preliminary analysis shows that information presented in ACCAP webinars has been used in developing training materials, conservation planning, private consulting, and climate literacy efforts across Alaska.

6.3.4 Network processes

Participants and presenters in the webinar series reported that, in addition to learning about and applying information from climate-related research, webinars help them connect with new constituents or clients and learn about their priorities, concerns, and needs, and expand the networking capacity of participating organizations, especially those that host satellite hub sites. In this way, the ACCAP webinar series fosters new network connections and promotes further dialogue and network building.

A wide range of participants reported using ACCAP webinars as a means of building new connections and expanding their individual (or organizational) networks. For example, a state employee who was new to Alaska attended multiple webinars soon after starting her new job in order to, "keep informed" and "get to know my clients and stakeholders" (Participant 11, College/University, 2013).

A local media reporter stated that "it's great to find out who the experts are within the state of Alaska, to know there are those local experts, that one can go to if needed" (Participant 3, Media, 2013). A webinar speaker from with the USFWS explained,

> ... [the webinar] gave us access to a broader group of stakeholders than we may have gotten from a different venue or holding a webinar on our own, so it was great to have this connection with people who tie into the ACCAP webinars and are interested in learning more about different issues that are going on in Alaska that are related to climate change.
>
> *(Speaker 1, USFWS, 2013)*

Several participants described ways in which participation in the ACCAP webinar series plays a critical role in furthering their organization's mission, assisting them in accomplishing goals, and expanding their organization's networking capacity. For example, the KBRR includes webinar attendance at their satellite site in their annual stakeholder outreach reporting. As the coordinator of the KBRR webinar satellite site explains,

> One of the missions of the Coastal Training Program is to provide up-to date scientific information to coastal communities or decision makers and ... [hosting ACCAP webinars as a satellite site] is a great way for us to be able to tap into a larger network and offer up-to-date scientific information to the communities.
>
> *(Participant 2, KBRR, 2013)*

Similarly, the science coordinator for one of the USFWS Landscape Conservation Cooperatives (LCCs) in Alaska, underscored how presenting at an ACCAP webinar catalyzed her organization's outreach efforts that depend on developing and maintain partnerships and engaging public and private organizations to address landscape-scale challenges.

The science coordinator for one of the Alaska LCCs also underscored the way in which the ACCAP webinar allowed her to build a new connection with one of her indigenous constituents. She further explains how this one new connection led to additional planned conversations with the intention of continuing to strengthen the relationships between the LCCs and its indigenous constituents,

> How to engage aboriginal participation and how to incorporate cultural values into the LCC planning process ... is something we have struggled with ... But through follow-up conversations with the [webinar] participant, we've been able to identify various individuals that we would like to talk [with] about ... how to move forward.
>
> *(Speaker 1, USFWS, 2013)*

Collaborating relationships between scientists and stakeholders is the foundation of use-inspired science [16]. Participants and speakers reported ways in which ACCAP webinars link scientists with stakeholders, providing critical capacity for forging collaborative partnerships. In some cases, these are new relationships, in other cases they are re-introductions of past acquaintances. For example, a hydrologist at the APRFC reconnected with a scientist at the International Arctic Research Center (IARC) via her presentation in the webinar series and is now working closely with that scientist and her graduate student who is using the NOAA hydrologic model to incorporate satellite data as a source of information on precipitation. The APRFC hydrologist explained that,

> Establishing that closer relationship, not only gets us insight into what they're learning, but finds a way to apply that into operations.... I had met [IARC scientist] a dozen years earlier at an American Meteorological Association meeting. But we had not retouched until after that ACCAP presentation.
>
> *(Participant 4, APRFC, 2013)*

One of the participants in the webinar presented by the Alaska LCCs saw a direct application of his research to the LCC mission and was inspired to contact the speaker about his research. As a result, the LCC Steering Committee endorsed his proposal, building use-inspired science and meeting regional climate science management needs. The webinar speaker noted,

> ... had it not been for the webinar, he would not have made that connection.
>
> *(Speaker 1, USFWS, 2013)*

The above examples show how the webinar series is promoting dialogue and network building among scientists and stakeholders, and thus how the construction of knowledge-to-action networks can promote use-inspired science.

Webinar participants also report that the webinar series provides opportunities for co-workers in the same organization yet with different disciplinary specialties, a focal point for intra-agency and interdisciplinary dialogue. Such interdisciplinary and intra-agency connections are particularly well illustrated by impromptu satellite site participation:

> I work in a regional office, so it's an opportunity for folks from [different] disciplines to get together and watch it [the webinar]. I've been in the room with folks from our air quality division and maybe a couple of natural resource managers or cultural resource managers and so as we're colleagues that know each other anyway because we work on other climate-change related topics, but in many cases, the topic or the speaker is something germane to what we're working on, so bringing us together in a room and listening to it we can have a conversation around the presentation itself and sometimes it inspires more conversation once the presentation is finished.
>
> *(Participant 6, NPS, 2013)*

As noted above, ACCAP webinars draw local participants to satellite hub viewing sites. Post-webinar discussions often ensue furthering local dialogue about climate change issues and promoting local network building. The following example is from the Kachemak Bay Research Reserve; however, similar discussions are reported at other satellite sites.

> Following the webinar we usually take 15-20 minutes to just debrief on the webinar, share our thoughts on the information that was learned [and share] any ideas on how the information could be applied to our work, thoughts about trying to follow up with that speaker. .... It's just the idea of sharing and seeing what new information was learned and how that information could be applied for the work that they [community members attending satellite site viewing] are doing.
>
> *(Participant 2, KBRR, 2013)*

Thus, satellite hub sites in particular, bring together participants that represent diverse community interests and needs. The webinars serve as a focal point for cross-cultural networking, for linking university scientists with agency resource managers, and for dialogue and continued networking on a multiple levels.

6.4 Discussion

There is mounting need to improve the accessibility and facilitate the application of science in climate adaptation on multiple levels [17]. In the Arctic, the time is particularly ripe for a transition from knowledge to action [18]; and the need for networking and coordination is especially acute in Alaska [15]. The ACCAP Climate Webinar Series provides a platform, for disseminating scientific information, establishing new network connections, catalyzing regional climate outreach and communication, building cross-network and cross-cultural awareness, and linking knowledge to action. In this section, we discuss how the webinar series serves cross-scale linking functions and supports use-inspired science. We also discuss the diffuse network structure of the ACCAP webinar series and how this network is nested in other existing networks.

The webinar platform allows for and promotes cross-scale linking by creating a venue for participants with experience and jurisdiction on the local, state, national, and international levels to engage. Local networking is also promoted through both organized and ad hoc satellite viewing sites that allow for both pre- and post-event face-to-face discussion and dialogue. These satellite sites also provide access to the webinars to people who may not have individual internet access. Cross-scale linking and local networking are further enhanced by the engagement of diverse and geographically dispersed participants.

The webinars serve to bridge the gap across scales by bringing together people and organizations with local, regional, state, national, and at times international interests and perspectives in a virtual environment. The most direct example of bridging across multiple scales occurs during webinars that present findings and generate discussion of national-scale reports such as engagement sessions for the USGCRP NCA, the Integrated Arctic Management Report to the President by the Alaska Interagency Working Group, and the America's Climate Choices report. These webinars bring findings from national-scale reports to an Alaskan audience and, conversely, also bring local and regional concerns and issues in Alaska to the national audience, including those in the Washington D.C. region. Indeed, groups such as the Arctic Research Commission and the National Science Foundation have contacted ACCAP requesting to present in the ACCAP webinar series with the intention of soliciting local and regional input and exposure (Figure 6.3).

The significance of promoting cross-cultural awareness and linking federal agencies with the concerns, issues, and priorities of rural, indigenous communities in Alaska cannot be overstated. Over 250 Native tribes rely on lands that are now under federal jurisdiction to meet nutritional food

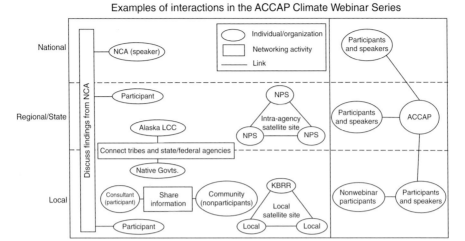

Figure 6.3 Schematic representing vertical and horizontal interplay that occurs in conjunction with ACCAP Climate Webinars. The vertical axis represents the scale at which information application or network interactions take place. The horizontal axis represents interactions within each level (i.e., local, regional/state, and national). The left side of the diagram illustrates specific examples of interactions and activities that occur, in part, as a result of the webinar series. The right side of the diagram shows a more general schema of ACCAP's networking interactions within the webinar series.

requirements and have traditional cultural ties to the land. Rural indigenous communities are heavily impacted by climate change and are not well served by existing federal institutional structures [19]. Enhancing local indigenous partnership, networking, and participation in climate-related science is an important path forward for climate adaptation in Alaska [20]. By facilitating network connections between federal agencies and indigenous communities, the ACCAP webinar series is one avenue for meeting this important need in Alaska.

At the same time, webinars serve to build and strengthen local network connections by convening diverse members in a community through organized satellite sites. Webinars also serve as a focal point for bringing together diverse employees in a single workplace, such as the ad hoc workplace viewing hubs at the NPS and APRFC. The post-webinar discussions that take place in these satellite sites build local capacity for collective information sharing and action and, in some cases, intra-agency networks for information exchange and problem solving. Together, the above examples highlight how ACCAP supports cross-scale linking functions and local networks, which provide a foundation for use-inspired science.

We identified multiple examples of ways in which the webinars served to build new relationships or invigorate existing partnering relationships between scientists (presenters) and stakeholders (participants), and led to collaborative research projects. These partnerships are both context- and content-specific in terms of the information needs of the stakeholders, the expertise of the scientists, and the spatial domain of analysis.

The cross-scale linking and local networking functions supported through ACCAP's webinar series is sustained by a fluid and dispersed network (see Chapter 9 and Chapter 11), rather than a more confined geographic community or a specific local or regional issue such as wildfire [3]. This fluid and dispersed network resulted in part from ACCAP's deliberate design of the webinar series to cover a broad spectrum of climate-related concerns across the state and reach new participants monthly. Such networking opportunities are critical for supporting knowledge-to-action networks in a rapidly transforming socio-ecological system. It is important to consider the dynamic nature of the network when designing and implementing program evaluation (see Section 6.5 and Chapter 10), as well as when documenting and evaluating the efficacy of adaptive actions that result from network interactions. These results demonstrate ways in which dispersed networks can foster climate adaptation.

The fluid and dispersed network we describe as being centered around the ACCAP webinar series, intersects with and is actually embedded within other intra-sector networks that play a critical role in supporting climate adaptation [4]. These networks will not only share some common members, but will also include other people and organizations. However, the knowledge-to-action network emanating from the ACCAP webinar series is dynamic and evolving, as illustrated in the development of formalized satellite viewing locations and changing network membership demonstrated by our high one-time participation rate.

6.5 Reflections on developing and maintaining knowledge-to-action networks

The goal of the webinar series since its inception has been to promote dialogue between scientists and stakeholders in order to encourage use-inspired science, identify research needs, and inform decision-making. We aim to create a forum for discussion and information exchange on the current state of knowledge about Alaska-specific aspects of climate change that is accessible to people statewide and identifies existing information gaps and how best to fill them. We wrap up this chapter by offering reflections on the process of implementing a webinar series and a few practical suggestions based on lessons learned over the past 6 years of hosting the Alaska Climate Webinar Series.

The webinar series explicitly promotes a dispersed, organically forming, dynamic network. Rather than being targeted toward a specific regional problem, set of policy decisions, or stakeholder community, this webinar series offers information and decision-support tools that address a broad diversity of topics, sectors, specific stakeholder interests, and sub-regions within Alaska. It is deliberately designed this way to meet the vast range of stakeholder demands that exist in Alaska and the Arctic due to rapid warming in northern latitudes and the need to design an effective response to a wide range of impacts already underway. The advantages of this approach are that it reaches a broad diversity of stakeholders and addresses a vast range of topics. As such, it builds an extensive stakeholder base, establishes widespread name recognition, and establishes ACCAP as the "go to place for climate information."[6] However, documenting and evaluating the evolution of a dispersed knowledge-to-action network and project outcomes is time-consuming and labor intensive.

Interviewees offered suggestions for improving the webinar series, most of which have been subsequently implemented in the ACCAP webinar series. These suggestions can also be helpful for other organizations working to implement a similar knowledge-to-action network. Suggestions included providing clear notice for how and where to participate in person; providing clear and easy access to a photo of the speaker, speaker contact information, and a video or similar archive of the presentation; providing short written summaries of the presentations; and grouping multiple presentations together into a half-day symposium. Interviewees also offered specific topics for future presentations, and suggested ways to improve the webinar software technology.

In critical reflection, more could have been done to document and track the actions that have resulted from webinar participation and the effectiveness of the webinar series in building a knowledge-to-action network and informing policy and decision-making. The webinar series was instituted at the end of ACCAP's first year of funding. In those early years, funding was limited and program focus was rooted in building stakeholder relationships, establishing trust, conducting research, and providing relevant and usable information. Consideration was given to the need for program and project evaluation, yet energy and attention was focused on building the stakeholder base rather than designing evaluation metrics.

---

[6]Stakeholder comment obtained from a 2010 web-based questionnaire. This questionnaire focused on obtaining evaluation feedback for various ACCAP programs.

This analysis suggests several avenues for future research on knowledge-to-action networks and on the ACCAP webinar series. For example, can social network analysis or other methods be used to investigate the cross- and inter-network information sharing and resultant actions? What would we need to know to determine how effectively the webinar series is nested within, and effectively builds upon, other established and emerging climate-related networks in Alaska and northern latitudes? Can more be done to more effectively integrate these networks? Is there a saturation point in cross-network integration that leads to diminishing effectiveness for climate adaptation? How does participant benefit or actions taken as a result of webinar participation vary with institutional affiliation and mission or with other variables such as a participant's role within the organization, the jurisdictional scale of decision-making (i.e., local, regional, state, national, international), the type of climate impact, the number of webinars attended, or the participant's location in the network structure?

Building and supporting a knowledge-to-action network requires understanding your constituents, or stakeholders, and anticipating their information needs. Consider, for example, what topics are seasonally relevant or of particular current interest to participants? How are they likely to use the information presented? What tips can be given to presenters to ensure that they reach a new and diverse audience? Webinar presenters must understand the application in the context of their research in order to bridge communication beyond raw research results with user application and actionable knowledge. In addition, network building involves surmounting logistical problems and maintaining a constant awareness of the limitations of the user group. For example, barriers such as time zones, internet access, bandwidth, or restrictions for software download on computers must be considered. Other boundary organizations that embark on creating, enhancing, or contributing to knowledge-to-action networks via webinars need to be advised to solicit consistent and frequent user feedback, institute systematic tracking of participation, and document the use of information in decision-making, and the distributed effects of webinar participation throughout the network. This should be done with due respect for the participants and an awareness of the dangers of "survey fatigue."

Such feedback could include understanding how participants evaluate the effectiveness or success of the webinar series, as these criteria may be different between the organizers and participants and also among participants. In the case of a webinar series, this could involve: (1) implementing on-line post-webinar questions for participants, (2) systematically archiving and recording unsolicited feedback about individual webinars, and (3) instituting bi-annual webinar participant interviews and web metrics analysis.

The goals of these evaluative methods would be to learn more about how webinar participation affects participant decisions, if and how webinar participation by both scientists and stakeholders promotes use-inspired science and how those collaborations evolve over time, and to track other outcomes of webinar participation such as network expansion and other features of information sharing.

6.6   Conclusion

Knowledge-to-action networks serve critical roles in climate adaptation and climate literacy. The ACCAP webinar series provides substantive information on a diversity of climate-related topics that is directly applied in trainings, public outreach, conservation planning, and consulting. With a diversity of topics, content, and audience, the webinars support network development and evolution by fostering new network connections, catalyzing regional climate outreach and communication efforts, and playing a keystone role in advancing network member success. The webinars further enhance the climate science and adaptation network in Alaska by building cross-network and cross-cultural awareness, and promoting science/stakeholder partnerships, and interdisciplinary, intra-agency, and local connections.

Through information exchange and network building, the webinar series supports the dissemination and application of information as well as the linking and integration of network members across local, regional, and national scales. As a regional platform with a trusted reputation, ACCAP webinars bring national-scale reports to the local level and local concerns to federal public servants. The webinars bridge geographic and cultural divides as they raise awareness of local indigenous concerns and issues for those working at regional, state, and national scales. Satellite sites where participants gather locally bring together communities and co-workers for additional post-webinar dialogue and discussion. By creating a virtual space for a diverse audience to convene around scientific topics, the webinars promote new and reinvigorated relationships between scientists, community members, representatives from federal agencies, the news media, and other boundary organizations. This generates opportunities for partnerships in research and application that provide the foundation for usable science.

Lessons learned that can inform others who engage webinars or other technologies in advancing a climate science and adaptation knowledge-to-action network are both contextual and procedural. Know your audience. Anticipate their interests, concerns, and technological constraints, and verify these assumptions by soliciting direct feedback, both formal and informal. Maintain a vision of your goals while remaining flexible to adjust to new opportunities and new technologies. Monitor your network's growth and document the interactions and outcomes that result from your program or activity.

Acknowledgments

This work was supported by the NOAA's Climate Program Office through Grant NA11OAR4310141 with the ACCAP at the University of Alaska at Fairbanks. The authors thank the webinar participants for their participation in this research. We acknowledge Karen Taylor for design and data collection in the 2010 webinar evaluation and Anna Schemper for interview transcription and review. The authors thank the reviewers for helpful comments. Any errors remain our own.

References

1 National Research Council (2005) Cash, D.W. & Buizer, J. (eds). Roundtable on Science and Technology for Sustainability. Policy and Global Affairs *Knowledge–Action Systems for Seasonal-to-Interannual Climate Forecasting: Summary of a Workshop*. The National Academies Press, Washington, DC.

2 Markon, C.J., Trainor, S.F. & Chapin, F.S. (2012) The United States national climate assessment – Alaska technical regional report. *U.S. Geological Survey Circular*, **1379**, 148.

3 Owen, G., McLeod, J.D., Kolden, C.A. *et al.* (2012) Wildfire management and forecasting fire potential: the role of climate information and social networks in the southwest United States. *Weather, Climate, and Society*, **4**, 90–102.

4 Dow, K., Haywood, B., Kettle, N. *et al.* (2013) The role of ad hoc networks in supporting climate change adaptation: a case study from the southeastern United States. *Regional Environmental Change*, **13**, 1235. doi:10.1007/s10113-013-0440-8

5 Armitage, D., Berkes, F., Dale, A. *et al.* (2011) Co-management and the co-production of knowledge: learning to adapt in Canada's Arctic. *Global Environmental Change*, **21**, 995–1004.

6 Cash, D.W., Clark, W.C., Alcock, F. *et al.* (2003) Knowledge systems for sustainable development. *Proceedings of the National Academy of Sciences*, **100** (**14**), 8086–8091.

7 Juhola, S. & Westeroff, L. (2011) Challenges of adaptation to climate change across multiple scales: a case study of network governance in two European countries. *Environmental Science and Policy*, **14** (**3**), 239–247.

8 Bodin, O. & Crona, B.I. (2009) The role of social networks in natural resource governance: what relational patterns make a difference. *Global Environmental Change*, **19**, 366–374.

9 Pahl-Wostl, C. (2009) A conceptual framework for analyzing adaptive capacity and multi-level learning processes in resource governance regimes. *Global Environmental Change*, **19**, 354–365.

10 National Research Council (2009) *Informing Decisions in a Changing Climate*. The National Academies Press, Washington, DC.

11 French, S. & Geldermann, J. (2005) The varied contexts of environmental decision problems and their implications for decision support. *Environmental Science and Policy*, **8**, 378–391.

12 Lackstrom, K., Kettle, N., Haywood, B. *et al.* (2014) Climate-sensitive decisions and time frames: a cross-sectoral analysis of information pathways in the Carolinas. *Weather, Climate, and Society,* **6** (**2**), 238–252.

13 Kasperson, R.E. & Berberian, M. (2011) *Integrating Science and Policy: Vulnerability and Resilience in Global Environmental Change.* Earthscan, London.

14 Dilling, L. & Lemos, M.C. (2011) Creating usable science: opportunities and constraints for climate knowledge use and their implications for science policy. *Global Environmental Change,* **21**, 680–689.

15 State of Alaska. (2011) *Adaptation Advisory Group of the Governor's Sub-Cabinet on Climate Change.* [cited]. Available from http://www.climatechange.alaska.gov/aag/aag.htm [accessed on 3 September 2015].

16 Feldman, D.L. & Ingram, H.M. (2009) Making science useful to decision makers: climate forecasts, water management, and knowledge networks. *Weather, Climate, and Society,* **1**, 9–21.

17 USGCRP (2012) *The National Global Change Research Plan 2012–2021: A Strategic Plan for the U.S. Global Change Research Program.* U.S. Global Change Research Program, Washington, DC, pp. 132.

18 USARC. (2013) *Report on the Goals and Objectives for Arctic Research 2013–2014,* pp. 24. US Arctic Research Commission.

19 GAO (2009) *Alaska Native Villages: Limited Progress has been Made on Relocating Villages Threatened by Flooding and Erosion.* US Government Accountability Office, Washington DC, pp. 49.

20 Knapp, C. & Trainor, S.F. (2013) Adapting science to a warming world. *Global Environmental Change,* **23**, 1296–1306.

SECTION III
Innovating services

The National Research Council defines climate services as "a mechanism to identify, produce, and deliver authoritative and timely information about climate variations and trends and their impacts on built, social-human, and natural systems on regional, national, and global scales to support decision making." Regional Integrated Sciences and Assessments (RISA) teams have made significant contributions to the literature on climate services including this definition. They have been both contributors to and test beds for concepts articulated in multiple NRC reports, for example, "Informing Decisions in a Changing Climate" [1], and "Americas Climate Choices" [2]. Innovating services examines the challenges of developing and implementing climate services.

In the debate around climate and weather services, a central concept is Research to Operations or Research to Applications, where scientific results produced in a research environment can be tailored to tools or products readily provided by operational offices of the federal government or private sector. Strategies abound for overcoming the "valley of death" [3], a process, institutional, resource, and skill barrier that exists between the opportunities created in research and the realization of a service product or tool. However, weather services benefit from greater predictability. Even seasonal to interannual climate predictions are more uncertain than weather predictions, making the development of a tailored product more difficult. In addition, climate vulnerabilities challenge decision-making over longer time scales and larger areas. Climate can impact operational or adaptive decisions over seasons, as well as near- and long-term planning decisions (e.g., reservoir releases vs investment in dams). Climate patterns can also impact areas outside of an organization's immediate jurisdiction, which still have an impact on the managed resource (e.g., availability of freshwater flows to downstream estuaries). Finally, because of our cognitive biases, we face many challenges presented by climate patterns.

Climate in Context: Science and Society Partnering for Adaptation, First Edition.
Edited by Adam S. Parris, Gregg M. Garfin, Kirstin Dow, Ryan Meyer, and Sarah L. Close.
© 2016 John Wiley & Sons, Ltd. Published 2016 by John Wiley & Sons, Ltd.

As Pulwarty *et al.* [4] note: "RISAs are experiments in the design and implementation of climate and environmental services. They are not the service itself." The chapters on Innovating services demonstrate that successful services rely on the integration of factors mentioned in previous chapters, such as understanding decision context and user needs, translating complex information into language that can be easily understood, and convening and facilitating processes to establish effective learning networks. While it is now abundantly clear that implementing services is not as simple as the oft-cited example of heaving data and information onto a loading dock [5], the authors contributing to Innovating services note specific technical and communication challenges associated with transferring programming code, and maintaining scientific consensus processes for producing services.

The case studies highlight challenges in developing the processes and products, as well as the challenges in communicating information to broad audiences of decision-makers and concerned citizens. The first chapter, by Garfin *et al.*, describes the specific example of developing and implementing forecasts of fire potential, informed by seasonal climate outlooks, to foster pro-active resource allocation in advance of upcoming fire seasons. They show the efficacy of developing temporary institutions for forecast experimentation and to build the capacity for applying climate knowledge to fire problems. In the second chapter, Carbone *et al.* document the critical steps in scaling up and extending a region-specific experimental decision-support product, the Dynamic Drought Information Tool, to reach diverse users across many states. They point to the advantages of collaborative software development, using formal product development protocols and guided by a technology liaison, as a means of generalizing online services; they suggest that this collaborative development process is an essential part of science translation. The final chapter on innovating services, by Shafer *et al.*, documents a rapid response to provide information services during the devastating 2011 drought in the South Central Plains. The authors demonstrate how multiple partners and multiple communications pathways (e.g., workshops, webinars, dialogues) reinforced each other, to sustain the multi-directional communication needed to meet decision-maker needs, and to engender trust in the information provided through the service.

References

1 National Research Council (NRC) (2009) *Informing Decisions in a Changing Climate*. The National Academies Press, Washington, DC.
2 National Research Council (NRC) (2011) *America's Climate Choices*. National Academies Press, Washington, DC.

3 National Research Council. (2000) *From Research to Operations in Weather Satellites and Numerical Weather Prediction: Crossing the Valley of Death*. Board on Atmospheric Sciences, pp. 80. National Academies Press, Washington, DC. http://www.nap.edu/catalog/9948.html.

4 Pulwarty, R.S., Simpson, C.F. & Nierenberg, C.R. (2009) The Regional Integrated Sciences and Assessments (RISA) program: crafting effective assessments for the long haul. In: Knight, C.G. & Jäger, J. (eds), *Integrated Regional Assessment of Global Climate Change*. Cambridge University Press, Cambridge, UK, pp. 367–393.

5 Cash, D.W., Borck, J.C. & Patt, A.G. (2006) Countering the loading dock approach to linking science and decision making: comparative analysis of El Niño/Southern Oscillation (ENSO) forecasting systems. *Science Technology and Human Values*, **31** (**4**), 465–494.

CHAPTER 7

The making of national seasonal wildfire outlooks

Gregg Garfin[1], Timothy J. Brown[2], Tom Wordell[3] and Ed Delgado[4]

[1] Climate Assessment for the Southwest (CLIMAS), School of Natural Resources and the Environment, Institute of the Environment, University of Arizona, 1064 E. Lowell St., Tucson, AZ 85721, USA

[2] California and Nevada Applications Program (CNAP), Desert Research Institute, 2215 Raggio Parkway, Reno, NV 89511, USA

[3] (Retired) USDA Forest Service, National Interagency Coordination Center, National Interagency Fire Center, 3833 S. Development Ave., Boise, ID 83705, USA

[4] Bureau of Land Management (BLM), National Interagency Coordination Center, National Interagency Fire Center, 3833 S. Development Ave., Boise ID 83705 USA

7.1 Introduction

Since at least 350 million years ago, fire and climate together have shaped landscapes and ecosystems [1]. Paleofire and paleoclimate records demonstrate the influence of climate on vegetation and the spread of fire. In recent centuries, the influence of society on fire ignition, spread, and management policy has increased the complexity of factors in an already complex, coupled social–ecological system. Human settlements in the wildland–urban interface (WUI) and across varied landscapes have increased, steadily exposing many more people and structures to fire, and accelerating human-caused fire ignitions. The U.S. Forest Service spends over $1 billion annually on fire suppression, in addition to the spending of all the other federal, state, local, and non-governmental fire services. At time scales of seasons to years, climate variability continues to be a key factor in understanding and anticipating fire occurrence and spread, either directly, through fire-year temperature, precipitation and wind, or indirectly, by preconditioning ecosystem fuel load and moisture.

Scientific understanding of the relationships between seasonal-to-interannual climate factors and fire has helped in developing seasonal fire forecasts. However, developing the capacity to convey forecast skill, and to interpret and use seasonal climate forecasts, has proved to be a substantial

Climate in Context: Science and Society Partnering for Adaptation, First Edition.
Edited by Adam S. Parris, Gregg M. Garfin, Kirstin Dow, Ryan Meyer, and Sarah L. Close.
© 2016 John Wiley & Sons, Ltd. Published 2016 by John Wiley & Sons, Ltd.

challenge in developing operational forecast products for many sectors of society [2]. For RISAs and other initiatives to innovate use-inspired services, crossing this "valley of death" [3] has been a quest to find the holy grail. The paths to the grail are typically filled with stories of needs assessment, capacity building, iterative engagement, and matching information supply and demand [4–7]. In this chapter, we document the innovation of consensus climate-informed preseason fire forecasts. We also reflect on the elements involved in bridging the gap between research and application, including the institutional partnerships and knowledge exchanges needed to foster the adoption of climate information in the fire sector. Sustaining this kind of process is a key RISA challenge. We examine the development of institutions and communities of practice as mechanisms to sustain the process.

7.2   Challenges in innovating climate forecast services

The use of probabilistic seasonal climate forecasts in decision-making—a key focus of the RISA program from its earliest stages [8]—poses challenges beyond those mentioned in previous chapters in Climate in Context (e.g., see Understanding Context). These additional challenges include lack of understanding about the interpretation and use of probabilistic information, inadequate demonstration of forecast skill, mismatch between forecast skill and decision timing, conservatism and resistance to the use of new and/or off-the-shelf information, and lack of trust in the messenger or deliverer of the forecasts [9–11]. In addition, there is a tendency for consensus forecast processes, which blend models and expert judgment (such as the seasonal outlooks produced by the National Oceanic and Atmospheric Administration (NOAA) and Climate Prediction Center (CPC)), to be overconfident, muddled by the increased complexity of assimilating many different types of forecast information, and prone to group conformity [12]. Even if all of the ideal climate information and forecasts were readily available and had adequate skill, using climate information across scales, from the on-the-ground resource managers through national policy-makers, is not a trivial process. As many have pointed out, it is not a function of simply making information or decision-support tools available, with the hope that they will someday get used [7]. Moreover, applying probabilistic seasonal forecast information to the context of natural resource management increases the complexity of interpreting the forecast, due to the high variability, nonlinearity, and complexity of natural systems [9]. To fully integrate climate information and fire risk assessment, for example, it is necessary to determine if the current climate record supports planning assumptions; to determine how planning and operations can benefit from what is known about climate; to determine entry points for climate information across scales in policy, management and operational responses; and to establish effective pathways from policy through operations at multijurisdictional levels [13]. This integration also requires developing a strong emphasis on process, mutual learning, and institutional stability for outputs and products (e.g., [6,14,15]). Thus, the integration of climate information and fire risk assessment provided a key opportunity for RISA investment, and an opportunity well integrated with the RISA program's aim to learn lessons for climate services development and implementation [8].

7.3   The intersection of fire and climate in science and fire policy

Scientists have long known about indirect relationships between climate and wildland fires. These relationships are mediated through climatic influences on vegetation type, year-to-year vegetation preconditioning in dry and wet episodes, and average length of fire season [16]. The possibility of predicting fire activity, from preseason climate information and forecasts, is a recent innovation. This was pioneered by scientists whose investigations of the relationships between long-lived El Niño-Southern Oscillation (ENSO) episodes clarified both the mechanisms linking climate variability directly to fire activity and the spatial relationships essential to understanding geographic variations in climate-related fire risk (e.g., [17–19]) (Figure 7.1). Earlier, fire-prediction studies emphasized weather (e.g., [20]), the characteristics (e.g., wind, humidity, precipitation) and effects of which are immediate, well understood and closely related to fire suppression—a main policy goal in the United States. However, climate prediction has evolved due to improved understanding of climate dynamics and persistent ocean–atmosphere interactions (e.g., ENSO) and the development and availability of long-term high quality meteorological observations.

United States fire policy has also evolved, and recently more focus has been on climate variability and change, especially climate change and drought [21]. Historically, catastrophic fire events (considered catastrophic by size and complexity, economic losses, fatalities, or all of the above, e.g., the 1910 "Big Blowup"[1]) have triggered key policy changes. Fire protection

---

[1]In 1910, a series of fires in the Pacific Northwest and Northern Rockies states, often referred to as the Big Blowup, burned more than 3 million acres, killing 85 people. The 1910 fire year put fire issues in the public eye, and led to implementation of fire prevention and suppression policies in the U.S. http://www.foresthistory.org/ASPNET/Policy/Fire/FamousFires/1910Fires .aspx.

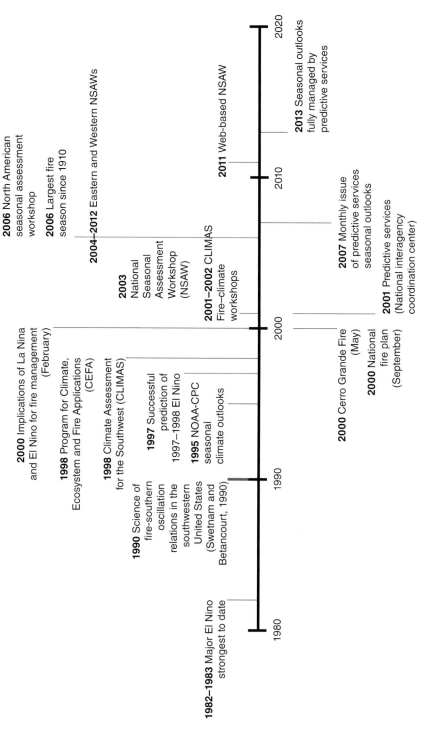

Figure 7.1 Timeline of key events pertaining to the development of seasonal fire potential outlooks.

has always been the chief emphasis, but as Pyne points out [22], policy has been a response to era-specific availability of land, money, human capacity, equipment, and information. Beginning with the Big Blowup, the focus of fire policy has shifted from economics, to systemized control, to prescription [22], reflecting perceptions of the chief fire problem and socioeconomic concerns of each era. Prior to 1988, severe drought and hot, dry winds were only implicitly related to large or catastrophic fire events. Following the 1988 Yellowstone fires, land management agencies increased their efforts to conduct prescribed burning in the western United States and to use prescribed fire to restore presettlement fuel loads. Post-Yellowstone initiatives began to explicitly emphasize the role of climate variability and change in fire occurrence [10]. Yet, climate, weather, fuels, fire, and society, are often considered independently, and have only recently been acknowledged as components of a complex and synergistic system.

The Beginnings of Predictive Services. Predictive Services (PS) is a slightly more than a decade-old institution in operational fire management in the United States (Figure 7.1). PS arose from a need to integrate weather, fuels, and suppression resource information into cohesive support products and services to help fire managers make better proactive and strategic resource allocation decisions. Prior to the early 2000s, the wildland fire management community was perceived as being more reactive than proactive, with the majority of fire management resources being allocated only as the need arose. Moreover, standard approaches to both fire suppression and pre-scribed burning did not consider the inter-relationships between weather, fuels, and resource requirements. Despite the development of important organizations, such as the NIFC and the Geographic Area Coordination Centers (GACCs), to coordinate logistics and mobilize resources for wildland fire emergencies, there were only ad hoc attempts to blend weather, fuels, and resource information into effective fire-management planning. These efforts were complicated by an increase in the number of hazardous fuel-treatment targets, declining budgets, more onerous and restrictive policies, regulatory guidelines, a large increase in the population living in the WUI, and a restructuring of the National Weather Service (NWS) Fire Weather Program [23]. The need for integrated fire management planning and decision tools grew as complicating factors multiplied and the consequences of resource allocation decisions became more far reaching.

In the 1990s, GACC fire intelligence coordinators, who were specialists in bringing together fire management information, produced outlooks that blended long-term weather outlooks with fuels and fire danger information. GACCs experimented with augmenting their outlooks, by collaborating with USFS and NWS fire meteorologists (e.g., in California), or hiring a meteorologist and fire weather program manager to bring expertise and help manage interdisciplinary information (e.g., in the Northwest Coordination Center). They solidified the approach of teaming fire meteorologists with intelligence coordinators, and established a model for a National Predictive Services Program. In the late 1990s, an internal agency proposal was submitted to the Secretary of Interior, with a goal of maximizing wildland firefighter safety and fire program efficiency, and ensuring the successful delivery and use of a wide-spectrum of fuels and fire weather information. This included the creation and operation of an interagency fire weather program within the federal wildland fire agencies, in order to provide a proactive role in management, operational activities, and oversight. Several alternatives, staffing scenarios, and funding levels were considered before this proposal was eventually adopted (Wordell, personal communication).

7.4   Fire–Climate workshops meet predictive services: setting the stage for seasonal fire outlooks

Swetnam and Betancourt examined fire scar records encompassing 400 years of fire history in the southwestern United States [18], and showed that the sequence of a wet El Niño winter followed by two dry winters, has a strong association with extensive wildfire occurrences. In the Southwest, La Niña usually results in a dry winter (Figure 7.1). The winter of 1997–1998 was wet and the summer of 1999 was very wet, allowing for substantial build-up of fine fuels—grasses and small shrubs—which foster the spread of fire, once an ignition has occurred. The winter of 1998–1999 was anomalously dry, and in the fall of 1999, the NOAA's CPC issued a 1999–2000 winter La Niña forecast with a high degree of confidence. If winter 1999–2000 was dry, as predicted, the sequence of wet and dry winters and their effects on fuels would strongly suggest high fire occurrence in the Southwest. Jim Brenner, a fire manager in Florida, also discovered a strong linkage between ENSO and fire in Florida [24]. Given ENSO's historic influence on fire across the southern tier of U.S. states, this meant an increased risk of a very active fire season.

With this information in hand, scientists from the Climate Assessment for the Southwest (CLIMAS) RISA, and departments at the University of Arizona convened a workshop, in February 2000, to bring fire managers and researchers together with climatologists, meteorologists, and climate impacts researchers to discuss the linkages between ENSO and wildfire regimes [25] (Figure 7.1). The exploratory workshop, armed with an information supply from the science community—a so-called "information push" [7]—aimed to bring together the producers and potential users of climate information, determine information needs and gaps, and begin to understand the decision contexts for entry of climate information into fire management decision-making. Participants represented federal and state agencies, including GACCs, federal research laboratories, tribes, and universities—including the recently established center for Climate, Ecosystems, and Fire Applications (CEFA, a partner in the California-Nevada Applications Program (CNAP) RISA) at the Desert Research Institute. Workshop participants recommended three things: (1) further research on fire-climate predictions and predictability, (2) development of a climate-related fire training program, as well as (3) the development of a pilot climate-fire outlook and assessment process for the Southwest. Workshop participants also identified potential uses for climate data and forecasts (Table 7.1).

As it turned out, the 2000 fire season was the most active on record at the time, with a fire somewhere in the country every day of the year. The 2000 Cerro Grande fire (Figure 7.1), which began as a prescribed fire, but

Table 7.1 Potential fire management uses of climate data and forecasts.

Assessment of approximately 40 million acres in National Forest System at elevated fire hazard
Firefighter safety
Resource losses
Escalating costs of fire suppression
Prevention and public planning
Setting priorities for allocation of firefighting resources at local, regional, and national scales
Multiagency coordination and decision-making and determination of preparedness levels
Long-range fire behavior prediction
Describing the level of uncertainty in fire behavior projections or model simulations
Supplemental and seasonal severity funding requests
Prescribed fire planning and priority-setting at the local, regional, and national scale
Operational decisions such as assessing the need to engage in mop-up activity, as well as determining line construction standards and firing sequences
Supporting landscape-level burn projects that continue over long time periods
Estimating the number of incidents over the course of a fire season
Selection of strategies and tactics, and in daily re-evaluations
Facilitate selection of strategies and cost estimates associated with wildland fire situation analysis development

Source: Adapted from [25].

escaped to destroy homes and threaten the city of Los Alamos, New Mexico, provided substantial impetus to include climate explicitly in fire policy. The postfire report noted that *moderate drought at least one year in the making* was not adequately considered in the prescribed fire planning and implementation [26]; strong winds, a seasonal characteristic of this region, *developed later, during the burn*, and fanned the flames into a significant wildfire event. Cerro Grande, and other large western United States fires, helped shape the USFS fire policy through the observation that fires are associated (1) directly with lack of precipitation, which causes droughts of various magnitudes, and (2) with predictable indirect climate controls on fuel build-up and desiccation. Impacts of the 2000 fire season prompted an infusion of nearly $1 billion in new funds for fire management, through the development and implementation of the National Fire Plan (NFP; Figure 7.1). By bringing together scientists and managers in the kind of knowledge exchange or learning-based mode championed by RISA [8], the workshop provided a foundation for the production of seasonal outlooks of significant fire potential. Prior to this workshop, the value of seasonal fire forecasts, and the specific users of forecasts were not apparent.

CLIMAS and CEFA partners were encouraged by the exchange of knowledge in 2000, and by wide support from the NOAA RISA program, the Joint Fire Science Program, the University of Arizona, and the USFS. CLIMAS and CEFA teamed up with collaborators from the University of Arizona's Laboratory of Tree-Ring Research and the NWS (Tucson Weather Forecast Office) to convene workshops in 2001 and 2002. The focus of the workshop process evolved from conveying the scientific basis for connections between prefire season climate and subsequent fire activity, to a focus on the specific application of information—climate and fire forecasts—for upcoming fire seasons. The 2001 and 2002 workshops also aimed to build capacity within the fire management community to understand and use climate forecasts (including a training session on the interpretation and potential uses of NOAA seasonal climate forecasts), and also within the climate science community to understand the management and societal contexts for enhancing fire preparedness [27,28].

Results from surveys and discussions conducted in association with the 2001 workshop highlighted potential uses of climate forecasts for preseason activities, such as planning, prioritizing allocation of firefighting resources, justifying additional funding needed to respond to above-normal fire danger, and management of prescribed fire programs [27]. Survey respondents ($n = 29$) noted that climate forecasts would be useful to "[i]ncrease lead time for severity funding requests (i.e., to get personnel on early when an early season is anticipated)" and to "[p]rovide heads up for managers when wildland fires may be used to accomplish resource objectives, and also when conditions are expected to be too harsh to allow fire use (i.e., all wildland fires need to be suppressed)." One respondent noted, "I believe we are talking about strategy and not necessarily tactics. Climate forecasts provide an opportunity to identify where additional resources may be needed and provide time to execute spending authority and prepare to move resources in a timely manner." Ninety percent of survey respondents agreed or strongly agreed that fire professionals need forecasts to include discussions on the climate mechanisms behind the forecast probabilities, and two-thirds of respondents indicated that "a low rate of incorrect forecasts," that is, a low false alarm rate, "is more important than a high rate of correct forecasts." The workshop revealed important constraints on the use of forecasts such as a lack of consideration, by the National Environmental Policy Act (NEPA), of climate and changes in fire regimes, and a fire management protocol that requires management decisions to be made on the basis of only the most recent 20 years' worth of data.

By the 2002 workshop, Predictive Services had evolved considerably. The NFP and infusion of new funds provided support for 20 fire meteorologists to join existing intelligence staff at the GACCs and the National Interagency Coordination Center (NICC) to form the National Predictive Services Program. A wildland fire analyst was added to the PS staff at NICC to enhance the analysis of fuels and fire behavior. This laid the foundation for PS units to serve as centers of expertise for the development and implementation of integrated fire tools and services, which could enhance the ability of managers to make sound, proactive, safe, and cost-effective decisions for *strategic planning and resource allocation*. Importantly, participants in the 2002 workshop included several of the newly hired PS meteorologists and intelligence personnel. In anticipation of PS participation in the 2002 workshop [28], CEFA coordinated the preparation of a consensus climate forecast for the 2002 fire season [29]. The consensus forecast process and workshop discussion took the first step toward addressing significant challenges in making a preseasonal fire outlook, by (1) explicitly articulating the climate factors associated with forecast probabilities, (2) explicitly articulating climate forecast confidence, (3) providing a forum for PS personnel to discuss climate forecasts directly with forecast producers, and (4) fostering discussions on integrated regional climate and fuels conditions for specific geographic areas.

A team from the Southwest GACC, NICC, and CEFA, issued an experimental Southwest seasonal fire danger forecast.[2] The team was emboldened

---

[2]The forecast was only issued on the internet (Southwest Coordination Center website). The authors were Richard Woolley, Jay Ellington (SW Coordination Center), Larry McCoy

by a combination of factors, including: confidence in the consensus forecast, the convergence of low fuel moisture and an early start to the 2002 fire season, and the climatological likelihood of strong wind events during the spring. Their experimental forecast included: an executive summary of key factors for the upcoming fire season, extensive descriptions of pre-existing climate-related factors (e.g., winter snowpack, precipitation, soil moisture), and fuel-related factors (e.g., herbaceous and large-fuel moisture), energy release component (a well-known parameter from the National Fire Danger Rating System; [30]), the 2002 workshop's consensus climate forecast, and three scenarios (most likely, best, and worst cases). This experimental forecast provided a template for the development of an operational product for the fire management community. Developing a reliable service required tackling unresolved issues, such as (1) the mismatch between the regional skill of NOAA-CPC seasonal precipitation outlooks and the need to forecast fire for all parts of the country, and (2) who would be responsible for the climate-informed seasonal fire outlooks?

7.5 National seasonal assessment workshops and the development of experimental seasonal fire potential outlooks

7.5.1 2003—The first national seasonal assessment workshop: laying the foundation

Following the 2002 workshop, Predictive Services joined with CLIMAS and CEFA, to develop a plan for a comprehensive national preseason fire outlook, informed by seasonal climate forecasts. Together, the organizations brought together human and financial resources from multiple federal agencies, a key aspect in forming an equitable and cohesive climate services partnership [15]. The partners (Figure 7.2) designed the NSAW to achieve the following goals: (1) foster communication and cooperation between climate forecasters and GACC personnel; (2) enhance collaboration and innovation; and (3) solidify a sustained partnership, in order to ensure continued knowledge exchange, service innovation, and product output. In addition, they aimed

---

(Kaibab National Forest), Tom Wordell (NICC), and Tim Brown (CEFA); they acknowledged the consensus climate forecast team: Tony Barnston (International Research Institute for Climate Prediction), John Roads (Scripps Institution of Oceanography), Rich Tinker (NOAA Climate Prediction Center), and Klaus Wolter (NOAA Climate Diagnostics Center and Western Water Assessment RISA).

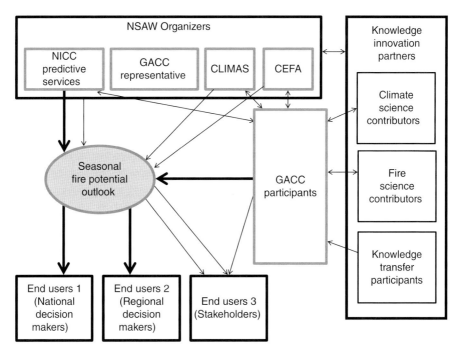

Figure 7.2 Partnerships, information, and communication flows contributing to the development of seasonal fire potential outlooks. Thicker arrows indicate stronger information and communication flows.

to catalyze cooperation and discussion between the GACCs, in order to: (1) improve fire outlook "edge-matching" in adjacent regions, by sharing information about regional fuels and climate/weather patterns, and (2) enhance the capacity of the GACCs to share methods, techniques, and information useful for producing their own seasonal outlooks [31].

The partners committed themselves to sustained interactions, equality in partnership, and clear partnership responsibilities. PS guided the workshop concept development process, took the lead on communication with project advisors and participants, and took responsibility for the fuels assessment activity—PS served as champions (*sensu* Brooks [15]) for the climate service, within the fire management community; CLIMAS developed the formal concept proposal, presented the concept to participants, and took responsibility for workshop logistics; and CEFA led a consensus climate forecast and ensured that forecast information was communicated clearly to fire specialists. The workshop concept was reviewed by an advisory committee, including potential workshop participants, and then presented to the GACC professionals at a business meeting approximately 4 months prior to the

NSAW. Prior to the workshop, and with input from the GACCs, the partners developed standards and protocols for data, information, and the production of multitimescale fire danger outlooks [31].

The multiday workshops followed a simple structure (covered in detail in [31]): Day 1, information exchange; Day 2, forecast development; Day 3, forecast critique and report synthesis. On the first day, forecasters presented climate and fire predictions, based on outputs from dynamical and statistical models, historical probabilities of above- and below-average temperature and precipitation, given analogous configurations of the ocean–atmosphere system, and knowledge of the strengths and weaknesses of the models and historical data. Panel discussion allowed participants to clarify concerns about their regions of interest, and allowed forecasters to compare notes. Fuels experts, fire behavior analysts and intelligence specialists presented analyses of current fuels conditions, areas of special concern (such as hurricane blow-down or insect-driven stand mortality), and regional climate and weather conditions leading up to the workshop. Climate forecasters developed a consensus seasonal climate forecast, with the aim of presenting probabilistic forecast information for regions of the United States that official NOAA forecasts characterized as having equal chances (EC) of above-or-below average conditions.

On the second day, the participants were divided into work groups based on geographic area, and climate forecasters roved among the work groups to provide region-specific insights, discuss special concerns, and learn about fire prediction needs and decision-making constraints. The work groups concentrated on assessing climate and fuels information—including information via conference calls to regional fuels specialists—and combining information into coherent predictions. Multiple groups conferred to synchronize their fire potential forecasts at geographic area boundaries. On the third day, participants presented and analyzed forecasts. Workshop organizers garnered feedback on the process, recommendations for improvements, evaluated shared problems and lessons learned, and synthesized the information for the briefing paper and media engagements.

The key output was a preseason assessment and consensus forecast of *significant fire potential*, which is defined as the likelihood of a wildland fire event requiring mobilization of additional resources from outside the area in which the fire situation originates. The product is a color-coded national map (Figure 7.3), accompanied by a report detailing climate, fuels, and potential weather information pertinent to the upcoming fire season. The map does not predict where or how much fire will occur, but to what extent land management agencies are likely to be in need with respect to additional regional

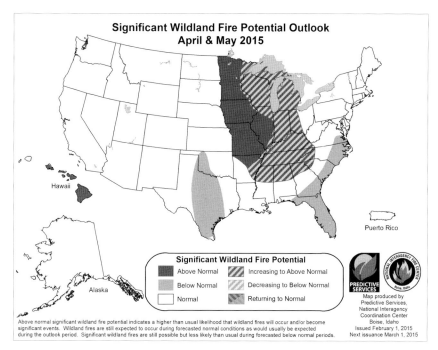

Figure 7.3 Example of a significant wildland fire potential (SWFP) outlook map issued in February 2015. Areas showing above-normal SWFP indicate a higher than usual likelihood that wildland fires will occur and/or become events requiring resources from outside the region. White areas indicate that wildland fires are expected to occur as would usually be expected; these areas will experience fires, perhaps with high amounts of area burned, but the region is capable of handling resource needs. Areas showing below-average SWFP indicate that significant wildland fires are still possible, but are less likely than usual during the forecasted period. Hatched areas (Increasing to Above Normal; Decreasing to Below Normal) depict change in potential throughout the forecast period. Outlook maps and discussions may be viewed at http://www.predictiveservices.nifc.gov/outlooks/outlooks.htm. Source: Predictive Services, National Interagency Fire Center, Boise, Idaho.

firefighting and fire management resources, during the fire season. This is a unique type of product, as it focuses on potential direct impacts on wildfire management agencies, rather than forecasts of weather, climate, or fire danger. Coproduction of the fire potential maps and report, by fire and climate experts from federal, state, and academic partners, lent legitimacy to the forecast; annual forecast evaluation, validation, and communication by PS and GACC personnel helped to bolster the credibility of the product [32]. Communication by PS to federal agency national managers, whose decisions, informed by the forecast, prioritized resource allocation, and support from NIFC leadership, ensured few organizational barriers to acceptance of the forecast for national and regional decision-making [32].

7.5.2   2004–2007—Refining the national seasonal assessment workshops

After the first NSAW, the partners convened two individual workshops per year (Figure 7.1), to account for winter-spring climate-fire seasonality in the eastern and southern geographic areas and the summer seasonality in the western geographic areas and Alaska; this set in motion participation by multiple RISA teams, including the Southeast Climate Consortium, the Alaska Center for Climate Assessment and Policy, and the Southern Climate Impacts and Planning Program. In response to requests from participants, we added workshop training sessions (e.g., [33]) to improve understanding of:

- climate phenomena, such as drought, teleconnections, and the North American Monsoon
- climate forecast procedures and techniques
- medium-range forecasts and resources for obtaining forecast information
- use of remotely sensed vegetation assessment products
- statistical techniques for forecast verification.

Other refinements included enhanced preworkshop preparation; priority access to national fire statistics databases; a session devoted to review of previous seasons' climate and fire potential forecasts; and, on-site participation by a GIS expert, to increase product accuracy. Workshop organizers garnered increased commitments from climate forecasters to consult with participants on region-specific issues, such as forecasts for Alaska.

In 2006, the partners added an experimental North American fire potential outlook, in collaboration with participants from Canadian and Mexican fire, forest, and meteorological services agencies (Figure 7.4).[3] Goals included: improving the flow of fire management information between the three countries, and improving the capacity of three nations to use climate forecasts and climate monitoring information in proactive fire management and resource allocation. This product is now part of operational preseason outlooks issued by PS.

There remain some specific challenges. First, the three countries have not yet agreed upon a definition for significant fire potential. For Canada and Mexico, the map is primarily depicting areas of anticipated above, below or normal fire activity, while the United States focuses on the significant wildfire potential definition. Second, it has been challenging to engage and coordinate individuals in each country, with the necessary expertise to

---

[3] Also in 2006, CEFA introduced the outlook concept to Australia, where, in the spring prior to southern Australia's fire season, a workshop was convened following a modified NSAW process [34].

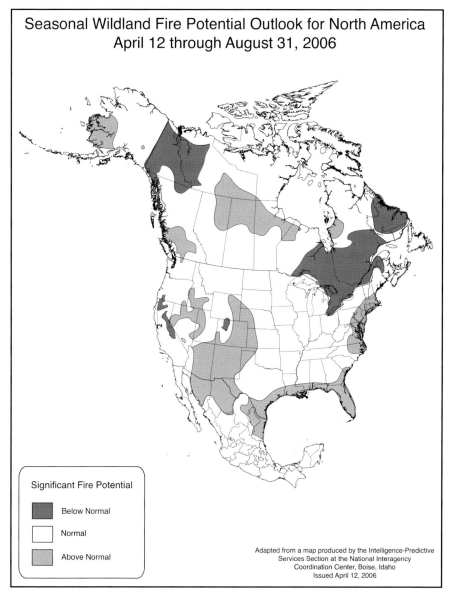

Figure 7.4 North American significant wildland fire potential outlook map, April 2006. The format is identical to the U.S. version (shown in Figure 7.3). Current outlook maps and discussions may be viewed at http://www.predictiveservices.nifc.gov/outlooks/outlooks.htm. Source: Predictive Services, National Interagency Fire Center, Boise, Idaho.

produce the outlook map. Third, fire seasons occur at opposite times of the year in Canada and Mexico, and in the United States fire can occur in at least one region any day of the year. To address seasonality, implementation of a more frequently coordinated operational product is being planned for 2015. The North American outlook is now coordinated with the North American Climate Services Partnership (NACSP http://cpo.noaa.gov/sites/cpo/Partnerships/International/NACSP/docs/EN-%20NACSP_Strategy_Final_2013-2017.pdf), to help facilitate the exchange of information, technology, and management practices, and the development and delivery of integrated climate services for North America.

7.6   Sustaining the process (2007–2014)

In 2007, in coordination with GACC personnel and climate service providers, PS began developing and operationalizing monthly outlooks covering the proximate month and season, with a 1-month lead time. During the spin-up of the *monthly seasonal* product, the Eastern and Western NSAWs continued through 2010, augmented by monthly coordination calls that followed a process similar to NSAW: GACCs generated preliminary outlooks and refined them, in consultation with outside experts. The shift from an experimental workshop-based process to an operational product generated some changes, as follows:

- With a structured operational process, GACC personnel developed the capacity to stay informed on the global and regional state of the climate, year-round, and increased their knowledge of climatology; two PS meteorologists now serve as climate experts within PS.
- The monthly updated product increased opportunities for PS meteorologists to develop trust with regional fire management entities through a continuous process of climate-related output, aimed at ongoing situational awareness and strategic planning.
- The combination of improved, easily accessible web-based meeting technologies, fiscal austerity prompted by the late-2000s economic recession, and increased confidence in the monthly seasonal call process stimulated PS to experiment with virtual meetings.

In January 2011, a teleconference-based *virtual* NSAW was convened (Figure 7.1), and it fostered further changes: (1) meetings were held in two half-day online sessions; (2) training was removed from the agenda, and replaced by brief talks to introduce new climate information and online tools. The virtual NSAW followed the same review, forecast, critique, and synthesis process as the in-person workshops. Decreased workshop time was accompanied by substantial GACC premeeting preparation, which left more time for discussion during the virtual workshop.

A key uncertainty leading up to the virtual NSAW was whether the new process would be as effective as in-person workshops. To gauge effectiveness, we garnered survey ($n = 14$) and focus group feedback from workshop participants. In general, participants were satisfied with exchange of information during the virtual NSAW. Over 80% of survey respondents indicated that communication of ideas and coordination with neighboring GACC regions were adequate, and in some cases exceeded the levels achieved during in-person workshops. We lack adequate data to attribute effectiveness to any single factor, such as ease of using webinar technology, or the experience gained in communicating via the monthly coordination calls. However, our survey result is in line with findings that the exchange of information in face-to-face meetings is no more effective than in virtual meetings [35]. Warkentin *et al.* note that face-to-face teams reported greater *satisfaction* with the group interaction process; similarly, we found that nearly all virtual NSAW participants pointed to the importance of in-person interactions, whether during the workshop or through informal conversations. One respondent mentioned "[p]lease don't write off in-person meetings entirely for the future. The expense is certainly justified, with the sidebar discussions that take place only in an in-person meeting." Warkentin *et al.* [35] conclude that virtual meetings may impede the development of a strong sense of group cohesion and reduce satisfaction with interactions and the effectiveness of information communicated. We come to similar conclusions in our evaluation of the 2011 virtual NSAW.

Based on participant feedback, the advantages of the virtual NSAW include time and cost savings, a more efficient and convenient process, and involvement by wider array of colleagues. Only 15% of the respondents noted dissatisfaction with the virtual meeting. Nevertheless, 65% of the survey respondents preferred a combined in-person and virtual meeting. Nearly 80% of the respondents indicated that participating from their office interfered with focused participation, due to interruptions by office colleagues, incoming phone calls, and a tendency to attempt multitasking. The 2012 virtual NSAW took even less time to conduct. We hypothesize that familiarity with the virtual process, improved preparation, and experience gained during the ongoing monthly coordination calls, contributed to greater efficiencies. Streamlining of the monthly seasonal coordination process, and a routine of attention to climate forecasts and information helped maintain the quality of the outlooks, despite the loss of in-person interactions.

Looking ahead. Climate-informed fire potential forecasts are now firmly established as a service to managers (Figure 7.1). The increasing confluence of people and fire, primarily through the wildland-urban and expanding exurban interfaces, will continue to be a focus of fire management and policy, and numerous scientific studies project an increase in fire activity compared to recent historical baselines (e.g., [36]). Proactive planning and decision-making, facilitated by climate-informed fire potential outlooks, are essential to the success of key policies, such as the National Cohesive Wildland Fire Management Strategy (http://www.forestsandrangelands .gov/strategy). Proactive planning also contributes to long-term strategies embraced by fire scholars, such as changing the combustibility of the land, in order to replace catastrophic stand-replacing wildfires with tame fires [1].

PS meteorologists, employees of five federal agencies dedicated to serving the fire management community (USDA Forest Service, Bureau of Indian Affairs, Bureau of Land Management, National Park Service, and U.S. Fish and Wildlife Service) represent geographic regions covering the entire United States. This diversity brings varied perspectives to foster learning, and to address a spectrum of information needs. PS meteorologists are open to learning new information and tools, as demonstrated by NSAW training sessions and other PS activities. Yet, most PS meteorologists lack the capacity, the statistical and dynamical modeling knowledge base, and the access to the latest scientific literature, which limits their ability to acquire cutting edge knowledge and translate it into practice. Improving climate science and forecasts will necessitate further knowledge exchange and training, in order to incorporate scientific innovations into practice. As PS lacks NSAW-based training, it will need to seek opportunities for training, especially in the areas of formal forecast validation, verification, and evaluation. Verification is essential for increasing the credibility of the outlooks, and evaluation is needed to ensure funding, to demonstrate the worth of incorporating new techniques and knowledge, and to sustain the process and the product.

To further strengthen the utility and usability of seasonal fire outlooks, we note the following additional challenges:

- The definition of significant wildfire potential is not sufficiently distinct to validate forecasts, because many fires require outside mobilization response by the very nature of the protocols of the firefighting community (i.e., the Incident Command System). PS is currently reviewing the need for a revised definition.
- It is not trivial to convert resource allocation and movement data into baseline information for quantitative forecast verification, and PS lacks the detailed information needed to develop a historical baseline. We offer a gap-bridging solution: systematically verify quantitative inputs to the

outlooks (e.g., climate forecast skill, fuel dryness levels, analog circulation patterns).

- Currently, the outlooks are deterministic. Probabilistic forecasts would allow for an indication of uncertainty, and the development of objectively derived guidance that could then be modified based upon forecaster knowledge—similar to NOAA-CPC model guidance that local NWS forecasters then modify as needed.
- There is a need for more and better documentation of the economic and planning value of seasonal outlooks, and their distribution and use, which would help to better understand if the outlook product is sufficiently meeting manager needs.

7.7   Measuring outcomes

In this section, we measure outcomes, post-hoc, through the following three evaluations: (1) validation of the NSAW forecasts, (2) assessment of the ways in which fire managers incorporate forecast information into management practices in the Southwest, and (3) a qualitative retrospective assessment of the NSAW process, using metrics from three sources [5,37,38].

Validating NSAW forecasts. Validating NSAW forecasts is exceptionally challenging, because the most obvious metrics of the preseason fire potential forecasts, namely, fire occurrence and area burned are mediated by lightning, human-caused fire starts, and management decisions. For example, the apparent mismatch between a forecast for exceptionally high fire potential and a fire season with relatively few acres burned may be the result of proactive management choices, such as wilderness closures and camping restrictions, or a single weather event that substantially increases fuel moisture. While there is ample quantitative information on forecast elements such as temperature, precipitation, fuel moisture, and soil moisture, verifying the movement of national fire resources is much more difficult, and attributing fire behavior, number of occurrences, or acres burned to climate conditions or management actions has proved elusive. NICC developed a semiquantitative experimental validation, by mapping the locations of large fires onto the preseason above- and below-normal forecast regions (not shown), but this only partially accounts for the multitude of factors mediating resource needs. Thus, developing satisfying forecast validation metrics is a key, ongoing, research need.

Assessing forecast use. Owen *et al.* [32] identified numerous outlook uses, through surveys and interviews conducted with a diverse group of fire managers in the Southwest Geographic Area (comprising Arizona, New Mexico, and a portion of western Texas) and a social network analysis to examine information dissemination. Their results confirm that the forecasts are most important for preseason planning, such as prepositioning resources, and planning prescribed burns. Some fire managers attach the outlooks to requests for additional suppression resources beyond what is locally and regionally available (e.g., increased firefighting staff, funding for air support, fire prevention activities)—a practice also used by national fire program leaders to coordinate with GACCs on setting budgets and hiring additional staff and other resources [32]. Operational use is most frequently deployed for forecasts of above-average fire potential. Moreover, social network analysis revealed that PS meteorologists have become a trusted source of information, and that regional fire managers prefer obtaining information directly from the PS meteorologists, rather than by gleaning information from online products. Fire managers socialize the forecasts, by using the forecasts as a basis for discussing the upcoming season with colleagues via conference calls, meetings, and daily briefing sessions. The forecasts are also used by the media to raise public awareness and firefighting situational awareness, when an active season is forecast. This regional study points to the need for a comprehensive nation-wide evaluation of NSAW forecast use.

Retrospective assessment. To further evaluate the NSAW process, we assessed the evidence from workshop proceedings and conversations with workshop organizers, using a combination of criteria, anchored by National Research Council metrics (Table 7.2). In general, we found that the production of preseason fire potential outlooks is characterized by a representative, coproduced process, with contributions from resources and staff time by RISA and federal agency partners, and outputs communicated to a broad range of stakeholders. The most substantial outcomes of the process are the operationalization of the forecasts, the demand for briefings, the streamlining of the forecast process, and the extension of the process to other countries.

Outlook users intensify the reach of information through social networks [32], and the demand for outlooks has increased in response to agency requests. Over a 12-year period (2002–2013), the outlook has transitioned from an experimental workshop pilot to a fully operational product, used by the highest levels of Federal agency fire management (e.g., [39]). We speculate that outlook use and increased demand for information is due to the trust of fire managers gained by Predictive Services meteorologists through their multiple timely information products and forecasts, and through the regular outreach interactions and activities that they have developed [32]. However, we note that the NSAW process has not been evaluated by peer-review, except through the work of Owen *et al.* [32], and through federal agency reports (e.g., [2]).

Table 7.2 Retrospective assessment of the national seasonal assessment workshop process.

NSAW attributes	Reference [37]	Reference [5]	Reference [38]
Process comanaged by NICC and RISA partners; adaptive learning and sharing of techniques; workshop organization that is responsive to participant needs and requests; decreased time necessary to develop forecast reports, as a result of improved preparation and enhanced understanding of workshop process	Number and frequency of quality scientist–stakeholder interactions; evidence of team integration	Process was representative; process was credible; solutions were implementable	Process metrics measure the course of action taken to achieve a goal: planning, strategy, leadership, promoting partnership
Funding commitment from partner agencies; increased commitment by climate forecast agencies to provide training, tailor forecasts, interact with participants	No equivalent metric	Stakeholders invested staff time or funding; costs and benefits were equitably distributed	Input metrics are tangible quantities put into a process: intellectual foundation, commitment of resources, leveraging of existing resources
Stakeholders acknowledge use of information for preseason resource allocation and funding decisions	No equivalent metric	No equivalent metric	Output metrics measure services delivered: peer-reviewed broadly accessible results, stakeholder judgment of results for decision-making, communication of results to an appropriate range of stakeholders

(continued overleaf)

Table 7.2 (continued)

NSAW attributes	Reference [37]	Reference [5]	Reference [38]
Earlier and more frequent forecast requests to NICC Predictive Services and GACC units; improved coordination between NSAW participants and contributing federal and state agencies; enhanced understanding of forecasts and workshop process); improved flow of information between GACCs; transition of process from annual experimental to monthly operational forecasts; transfer of the NSAW process to other countries; increased capacity of climate community to tailor NSAW-specific information and decision-support tools	Science changes in response to user inputs; demand for climate information or briefings increases	Participants modify behavior; participants initiate subsequent contacts; relationships are sustained and extend beyond individuals to institutions; project took on a life of its own (became self-sustaining); project built capacity; stakeholders claimed partial ownership of final products	Outcome metrics are defined by the use and influence of outcomes: identification of uncertainties, improved understanding, transition of results to operational use, creation of institutions and capacity
Use of information in resource allocation decisions (especially at the national level)	Regional vulnerability reduced	Quality of life or economic conditions improved	Impact metrics measure the long-term environmental or societal consequences of an outcome: improved decision-making, protecting life and property, reducing societal vulnerability, increased public understanding

Sustaining the process for more than 10 years has created a new community of practice, and new capacity to use climate information in decision-making. We have shown evidence of incremental improvement of established routines, since the establishment of the NSAW process, through streamlining of the process—in the parlance of Pahl-Wostl [40], single-loop learning. We argue that this community has also established double-loop learning, through experimentation with innovative approaches, such as incorporation of experimental climate forecasts and information, development of internal forecasts through PS climate experts, and through the adoption of an entirely new operational approach during the course of more than a decade of experimentation. However, we do not see evidence that the community has called into question guiding assumptions or changed underlying paradigms and norms, which are facets of double- and triple-loop learning, respectively.

7.8   Implications for innovating services

Through our retrospective evaluation, we identified five factors that supported the codevelopment of operational preseason fire potential forecasts by PS, CLIMAS, and CEFA/CNAP: the development of new institutions, focus on a broad geography, a participatory deliberative process, inclusion of many partners, and sustained engagement. First, the initiative was supported by the development of two temporary institutions, the 2000–2002 Fire–Climate Workshops, and the 2003–2012 NSAWs. These institutions helped RISAs identify early adopters and partners, fostered and sustained connections between producers and users of climate information, laid the foundation for codeveloping an experimental service, fostered learning, and helped transition an experiment into an operational service. The novelty of these institutions coincided with the development of two new organizations: RISA and Predictive Services. This allowed for experimentation where the lines between scientists, practitioners, and disciplinary boundaries were blurred [41]; multidirectional knowledge flow was fostered; and, through sustained and iterative interaction, new social capital and trust between disciplines were developed [6]. We speculate that, because these were new organizations, there was a greater openness to new approaches, and a need to work together to define measures of added value, process and product efficacy, and product validation. The new institutions provided critical support and a foundation for the coproduction of process, products, and knowledge.

Second, the process aimed at a large geography—areas of the contiguous United States with discernible and strong relationships between interannual climate variability and fire. This facilitated a rapid transition to working at the national level required by Predictive Services, which was the operational partner. The extensive geography fostered participation and contributions from multiple RISAs, NOAA laboratories, and Regional Climate Centers, and, thus, hastened widespread knowledge exchange, and rapidly built capacity for the use of climate information in decision-making. Working at the national scale shed light on multiple facets of seasonal fire prediction, such as the need for seasonal outlook products that address the likelihood of multiday blocking events, which have a pronounced impact on fuel conditions in the central and eastern United States.

Third, coproduction of the seasonal fire potential forecasts facilitated interactions between producers and users of climate knowledge in a structured, participatory deliberation with expert analysis [11], which promoted learning about climate forecasts and, to some degree, engendered confidence in the forecast products. Over many years, PS meteorologists embraced and incorporated NOAA-CPC probabilistic climate forecasts into fire potential outlooks, if (1) participating climatologists could adequately explain the climate system mechanisms that generated the predicted probability anomaly patterns, and (2) the forecasts were accompanied by the types of geopotential height maps commonly used by meteorologists. Nevertheless, many PS meteorologists prefer to use multiple lines of evidence, including NOAA-CPC probabilistic forecasts, and forecasts based on atmospheric circulation patterns from prior years with similar circulation patterns (so-called "analogue years")—which are easier to interpret. However, as highlighted by Van den Dool [42], there is considerable constraint in the skill of analogue forecasts, because they require a very large number of patterns to generate statistical confidence in the forecast.[4] Constructed analogues [42], used by NOAA-CPC, improve forecast skill, but the method has not caught on with PS meteorologists. This suggests that a new loop of learning is needed to promote the use of best practices [40]. We suggest that new methods could be delivered effectively by meteorologists trusted by the Predictive Services community. One possible approach is embedding scientists in PS offices. Alternatively, PS could hire GACC climatologists, with sufficient training in

---

[4]While this statement focuses on statistical accuracy, it is interesting to consider that decision-making relies on experiences, and it is easier to relate to historical patterns of an event than to a new pattern [43]. Thus, some decision makers may have more confidence in historical analogues than probabilistic forecast patterns as a predictor of current conditions, despite the low statistical power of analogues, due to their familiarity with historical analogues.

statistical and dynamical modeling methods, to work side-by-side with PS meteorologists.

Fourth, as the process progressed, it was important to develop a "big tent"—a diverse base of collaborators and partners from federal and state entities and universities (e.g., NOAA and USDA-FS labs, NWS offices, Regional Climate Centers)—to infuse new knowledge, expertise, and agency know-how. The strong existing interpersonal ties between workshop organizers and potential partners likely accelerated discipline-bridging collaborations and promoted trust in the process.

Finally, as noted by many (e.g., [15]), sustained engagement, iterative interactions through which process and knowledge can evolve, and secure funding accompanied by commitment from partners and their organizations helped the initiative to innovate services and bridge the gap between research and operations. The novelty of the PS concept, along with a willingness of individuals to learn, reinforced the workshop process. Over the years, as trust in PS increased, the GACC personnel valued the outlooks more, which led to refinements and operationalization of the product.

7.9 Conclusions

Innovating, implementing, and operationalizing climate services for decision-making communities—a process likened to crossing the valley of death—continues to challenge RISA and a new generation of initiatives, such as the Department of Interior Climate Science Centers and USDA Climate Hubs. In this chapter, we described how we overcame barriers noted by the National Research Council, to codevelop and implement climate-informed seasonal fire potential outlooks used by stakeholders as prominent as the Secretary of Agriculture and as commonplace as district fire management officers. We conclude by reflecting on the characteristics of this process, some serendipitous and some intentional, that could be of use to other climate service providers.

There was a certain serendipity in the timing of our efforts, beginning with a single knowledge-exchange workshop, and the confluence of events that lent gravitas, momentum, and determination to carry the process from innovation to operational use, namely: a series of years with fires of spectacular extent, the development and implementation of the NFP, public attention to the issue, and the continuity and expansion of the RISA program. The Predictive Services Program and RISA were well-matched partners, in part, because of the national and regional scales at which they operate. In addition, they were new organizations, which allowed sufficient flexibility to experiment.

NSAW organizers and collaborators had already established some personal relationships, which engendered trust and empowered champions within the partner organizations.

On the intentional side, at a sufficiently high level, both NOAA and the agencies comprising the NIFC were willing to focus efforts on addressing stakeholder needs, which facilitated participation from multiple branches within agencies (e.g., GACCs, NOAA and USDA-Forest Service research labs, BLM, NOAA-CPC, NWS). Commitment from high-levels, backed by sufficient funding for a sustained process, allowed organizers and on-the-ground participants sufficient time to develop new institutions, and gain familiarity with experimental processes, and to modify them. By inviting participation by a broad array of partners and collaborators, the organizers fostered the development of a community of practice. Simultaneously, the ongoing presence of extensive wildland fire reinforced the need for the experimental product to address resource allocation and preparedness decisions, and therefore maintained demand by the user community. Yet, the ongoing presence of extensive wildland fire underscores how far this climate service needs to go to achieve full success, by advancing the interplay between basic science, improved forecast products, use of those products, and the reduction of risk—the last of which will require substantial insight from the social sciences.

Acknowledgments

Funding to support the development of seasonal fire potential outlooks came from the NOAA Climate Program Office, USDA-Forest Service, Joint Fire Science Program, the Department of Interior agencies comprising the NIFC, and the University of Arizona. We thank the NOAA Earth System Research Laboratory for hosting the majority of Western NSAWs, and the National Conservation Training Center for hosting the Eastern NSAWs. We thank the NOAA CPC for donating climate forecaster time to attend every NSAW, and for contributing to operational fire potential outlooks. We also acknowledge the contributions of time and expertise by dozens of climate and fire science researchers and practitioners from federal agencies and their research laboratories, universities, state forestry, and natural resource departments, RISA teams (notably, the CNAP, the Western Water Assessment, and the Southeast Climate Consortium), and Mexican and Canadian federal agencies. We particularly thank the following individuals for getting the process off the ground, and for sustaining it over the years: Barbara Morehouse (University of Arizona, ret.), Rick Ochoa (BLM, ret.),

Tom Swetnam (University of Arizona, ret.), Klaus Wolter (NOAA-ESRL, CIRES), and Robyn Heffernan (NWS).

References

1 Scott, A.C., Bowman, D.M.J.S., Bond, W.J. *et al.* (2014) *Fire on Earth: An Introduction*, 1st pp. 413 edn. Wiley Blackwell, Hoboken, New Jersey, USA.

2 CCSP (2008) *Decision-Support Experiments and Evaluations using Seasonal-to-Interannual Forecasts and Observational Data: A Focus on Water Resources*. A Report by the U.S. Climate Change Science Program and the Subcommittee on Global Change Research N. Beller-Simms, H. Ingram, D. Feldman, N. Mantua, KL. Jacobs & A. Waple (eds), pp. 192. NOAA's National Climatic Data Center, Asheville, NC.

3 Barr, S.H., Baker, T. *et al.* (2009) Bridging the valley of death: lessons learned from 14 Years of commercialization of technology education. *Academy of Management Learning & Education*, **8** (**3**), 370–388.

4 Lemos, M.C. & Morehouse, B.J. (2005) The co-production of science and policy in integrated climate assessments. *Global Environmental Change*, **15**, 57–68.

5 Jacobs, K.L., Garfin, G.M. & Lenart, M. (2005) More than just talk: connecting science and decisionmaking. *Environment*, **47** (**9**), 6–22.

6 McNie, E.C. (2007) Reconciling the supply of scientific information with user demands: an analysis of the problem and review of the literature. *Environmental Science & Policy*, **10**, 17–38.

7 Dilling, L. & Lemos, M.C. (2011) Creating usable science: opportunities and constraints for climate knowledge use and their implications for science policy. *Global Environmental Change*, **21**, 680–689.

8 Pulwarty, R.S., Nierenberg, C. & Simpson, C. (2009) The Regional Integrated Sciences and Assessment (RISA) program: crafting effective assessments for the long haul. In: Knight, C.G. & Jäger, J. (eds), *Integrated Regional Assessment of Global Climate Change*. Cambridge University Press, Cambridge (UK), pp. 367–393.

9 Rayner, S., Lach, D. & Ingram, H. (2005) Weather forecasts are for wimps: why water resource managers do not use climate forecasts. *Climatic Change*, **69**, 197–227.

10 Corringham, T.W., Westerling, A.L. & Morehouse, B.J. (2008) Exploring use of climate information in wildland fire management: a decision calendar study. *Journal of Forestry*, **106**, 71–77.

11 Feldman, D.L. & Ingram, H.M. (2009) Making science useful to decision makers: climate forecasts, water management, and knowledge networks. *Weather Climate Society*, **1**, 9–21.

12 Nicholls, N. (1999) Cognitive illusions, heuristics and climate prediction. *Bulletin of the American Meteorological Society*, **80** (**7**), 1385–1397.

13 Brown, T.J. & Pulwarty, R.S. (2006) *A Fire Place for Climate: The Role and Use of Climate Information in Fire Management Policy*. Proceedings Fifth International Conference on Forest Fire Research, 27–30 November 2006, Portugal, 9 pp.

14 National Research Council (2009) *Informing Decisions in a Changing Climate* Panel on Strategies and Methods for Climate- Related Decision Support. Committee on the Human Dimensions of Global Change, Division of Behavioral and Social Sciences and Education, 188 pp. National Academies Press, Washington, DC.

15 Brooks, M.S. (2013) Accelerating innovation in climate services: The 3 E's for climate service providers. *Bulletin of the American Meteorological Society*, **94**, 807–819.

16 Schroeder MJ and Buck CC. (1970) Fire climate regions. Pages 196–220 (Chapter 12). In: *Fire Weather : A Guide for Application of Meteorological Information to Forest Fire Control Operations.* Agriculture Handbook 360. USDA Forest Service, Washington, DC.

17 Simard, A.J., Haines, D.A. & Main, W.A. (1985) Relations between El Nino/southern oscillation anomalies and wildland fire activity in the United States. *Agricultural and Forest Meteorology*, **36**, 93–104.

18 Swetnam, T.W. & Betancourt, J.L. (1990) Fire-southern oscillation relations in the southwestern United States. *Science*, **249**, 1017–1020.

19 Westerling, A.L., Brown, T.J., Gershunov, A. *et al.* (2003) Climate and wildfire in the western United States. *Bulletin of the American Meteorological Society*, **84**, 595–604.

20 Rothermel, R.C. (1983) *How to Predict the Spread and Intensity of Forest and Range Fires.* USDA Forest Service, Intermountain Forest and Range Experiment Station, General Technical Report INT-143, 161 pp.

21 USDA Forest Service Fire & Aviation Management and Department of the Interior Office of Wildland Fire (2014) *Quadrennial Fire Review – Subtask 2.2 Report Card.* USDA Forest Service Fire & Aviation Management and Department of the Interior Office of Wildland Fire, 71 pp. URL http://www.forestsandrangelands.gov/QFR/documents/QFR_Report_Card_04182014.pdf [accessed on 3 September 2015].

22 Pyne, S.J. (1997) *Fire in America: A Cultural History of Wildland and Rural Fire.* University of Washington Press, USA, pp. 654.

23 Winter, P.L. & Wordell, T.A. (2010) An evaluation of the Predictive Services Program. *Fire Management Today*, **69**, 28–33.

24 Brenner, J. (1991) Southern oscillation anomalies and their relationship to wildfire activities in Florida. *International Journal of Wildland Fire*, **1**, 73–78.

25 Morehouse, B.J. (2000) *The Implications of La Niña and El Niño for Fire Management.* Climate Assessment Project for the Southwest (CLIMAS) Workshop Proceedings, February 23–24, 2000, pp. 45. Tucson, AZ.

26 National Park Service (2001) *Cerro Grande Prescribed Fire: Board of Inquiry Final Report*, pp. 47. National Park Service, Washington, DC. URL http://www.fireleadership.gov/toolbox/staffride/downloads/lsr6/lsr6_Cerro_Grande_Board_of_Inquiry_Report_Feb_2001.pdf [accessed on August 23 2013].

27 Garfin, G.M. & Morehouse, B.J. (2001) *Facilitating Use of Climate Information for Wildfire Decision-Making in the U.S. Southwest.* American Meteorological Society, Fourth Symposium on Forest and Fire Meteorology, Reno, NV, November 13–15, 2001. pp. 116–122. American Meteorological Society, Boston. URL http://www.climas.arizona.edu/sites/default/files/pdfgarfin2001.pdf [accessed on 3 September 2015].

28 Garfin, G.M. & Morehouse, B.J. (2005) *2002 Fire in the West* Workshop proceedings, March 5–6, 2002. Institute for the Study of Planet Earth, Tucson, Arizona., pp. 83.

29 Brown, T.J., Barnston, A.G., Roads, J.O. *et al.* (2002) 2002 Seasonal Consensus Climate Forecasts for Wildland Fire Management. *Experimental Long-Lead Forecast Bulletin*, **11** (**March 2002**), 4.

30 Hardy, C.C. & Hardy, C.E. (2007) Fire danger rating in the United States of America: an evolution since 1916. *International Journal of Wildland Fire*, **16**, 217–231.

31 Garfin, G.M., Wordell, T., Brown, T.J. *et al.* (2003) *The 2003 National Seasonal Assessment Workshop: A Proactive Approach To Preseason Fire Danger Assessment.* Proceedings of the American Meteorological Society 5th Symposium on Fire and Forest Meteorology, Orlando, Florida, November, 2003. Extended Abstract J9.12, pp. 7. Boston, American Meteorological Society. URL https://ams.confex.com/ams/FIRE2003/webprogram/Paper71583.html [accessed on 3 September 2015].

32 Owen, G., McLeod, J.D., Kolden, C. *et al.* (2012) Wildfire management and forecasting fire potential: the roles of climate information and social networks in the southwest United States. *Weather, Climate and Society*, **4**, 90–102.

33 Lenart, M., Brown, T., Ochoa, R. *et al.* (2005) *National Seasonal Assessment Workshop: Western States & Alaska. Final Report*, May 2005, pp. 29. CLIMAS, Tucson, AZ.

34 Bushfire, C.R.C. (September 2006) Seasonal Bushfire assessment 2006–2007: Australian fire season outlook. *Fire Note*, (**5**), 2.

35 Warkentin, M.E., Sayeed, L. & Hightower, R. (1997) Virtual teams versus face-to-face teams: an exploratory study of a web-based conference system. *Decision Sciences*, **28**, 975–996.

36 Flannigan, M.D., Krawchuk, M.A., de Groot, W.J. *et al.* (2009) Implications of changing climate for global wildland fire. *International Journal of Wildland Fire*, **18** (**5**), 483–507.

37 Bales, R.C., Liverman, D.M. & Morehouse, B.J. (2004) Integrated assessment as a step toward reducing climate vulnerability in the southwestern United States. *Bulletin of the American Meteorological Society*, **85**, 1727–1734.

38 National Research Council (2005) *Thinking Strategically: The Appropriate Use of Metrics for the Climate Change Science Program*, pp. 150. National Academies Press, Washington, DC. http://www.nap.edu/catalog/11292.html

39 Newman, C. (2011) *Statement of Corbin L. Newman, Jr. Regional Forester, Southwestern Region U.S. Forest Service U.S. Department of Agriculture*, Before the Energy and Natural Resources Committee, U.S. Senate. April 27, 2011, 7 pp. URL http://www.energy.senate.gov/public/index.cfm/files/serve?File_id=987d8032-96e6-fb69-2f94-51030480b97e [accessed on February 15 2015].

40 Pahl-Wostl, C. (2009) A conceptual framework for analysing adaptive capacity and multi-level learning processes in resource governance regimes. *Global Environmental Change*, **19** (**3**), 354–365.

41 Rice, J.L., Woodhouse, C.A. & Lukas, J.J. (2009) Science and decision making: water management and Tree-Ring Data in the western United States. *Journal of the American Water Resources Association*, **45** (**5**), 1248–1259.

42 Van den Dool, H.M. (1994) Searching for analogues, how long must we wait? *Tellus*, **46**, 314–324.

43 Kahneman, D. (2011) *Thinking, Fast and Slow*. Farrar, Straus and Giroux, New York, pp. 499.

CHAPTER 8

Challenges, pitfalls, and lessons learned in developing a drought decision-support tool

Greg Carbone[1], Jinyoung Rhee[2], Kirstin Dow[1], Jay Fowler[1], Gregg Garfin[3], Holly Hartmann[4], Ellen Lay[4] and Art DeGaetano[5]

[1] Carolinas Integrated Sciences and Assessments RISA, Department of Geography, University of South Carolina, 709 Bull Street, Columbia, SC 29208, USA

[2] APEC Climate Center, 12, Centum 7-ro, Haeundae-gu Busan 48058, South Korea

[3] Climate Assessment for the Southwest (CLIMAS), School of Natural Resources and the Environment, Institute of the Environment, University of Arizona, Tucson, AZ 85721, USA

[4] University of Arizona, Tucson, AZ 85721, USA

[5] Earth and Atmospheric Science and Northeast Regional Climate Center, Cornell University, Ithaca, NY 14850, USA

8.1 Introduction

Drought is a slow onset climate-related hazard with a natural hazard component (lack of precipitation or water) and a social vulnerability component (e.g., lack of preparedness, or an insufficiently robust economy, infrastructure, laws, and policies). The United States suffers billions of dollars in drought losses each year, yet it lacks a national drought plan. In 2006, the President and Congress took one small step toward remedying this situation by passing the National Integrated Drought Information System Act of 2006 [1]. The goal of the Act is to improve the nation's capacity to proactively manage drought-related risks by providing affected communities with the best available information and tools to assess drought impacts and to improve the nation's ability to better prepare for and mitigate the effects of drought. Two of the key objectives stated in the NIDIS implementation plan include (1) creating an early drought warning system capable of providing accurate, timely, and integrated information on drought conditions and risks at spatial scales relevant to facilitate proactive responses and (2) providing interactive early warning information delivery systems for easily comprehensible and standardized products (such as databases, maps, forecasts, etc.) [2].

Climate in Context: Science and Society Partnering for Adaptation, First Edition.
Edited by Adam S. Parris, Gregg M. Garfin, Kirstin Dow, Ryan Meyer, and Sarah L. Close.
© 2016 John Wiley & Sons, Ltd. Published 2016 by John Wiley & Sons, Ltd.

Beginning in August 2007, in support of the NIDIS effort, the National Oceanic and Atmospheric Administration (NOAA) Regional Integrated Sciences and Assessments (RISA) program funded several projects to enhance stakeholder access to drought decision-support resources. We report here on one project that demonstrates the transferability, scalability, and expandability of region-specific decision-support tools by adapting the Carolinas Integrated Sciences and Assessments' (CISAs') Dynamic Drought Index Tool (DDIT) [3] for use in Arizona and New Mexico and in the 18 states represented by the Northeast and Southeast Regional Climate Centers (RCCs). This chapter summarizes the project process and lessons learned, including the implications for NIDIS in terms of public sector technology transfer, regional stakeholder needs and decision-support preferences in the natural resources and water management sectors, and enhancements to decision-support tools for effective use by stakeholders. In the context of our specific example, we explore two questions: What are the technical and institutional challenges associated with transferring and scaling an existing software tool? How can cooperation between institutions facilitate the transfer of software from research to operations? This chapter begins with an introduction to the drought tool and a review of literature pertaining to technology transfer and collaborative development. Then it documents the process of incorporating user needs into the evolving tool, highlights technical considerations associated with its expansion and scaling, and describes the team collaboration required for these tasks. Finally, we share lessons learned and prescribe recommendations based on our experience.

8.2 Background

8.2.1 The Dynamic Drought Index Tool

CISA's DDIT accommodates decision-makers who must consider drought in different physical, temporal, and political units and in the context of state and local ordinances. The tool was born out of crisis management during an extreme event (the 1998–2002 drought) and a regulatory requirement for multiyear Federal Energy Regulatory Commission (FERC) dam relicensing negotiations. The DDIT creates maps, graphs, and tables using a suite of drought measures at spatial and temporal scales selected by a user [3]. Originally developed for the Carolinas, the DDIT uses open architecture software to calculate drought indices from station data provided by the Applied Climate Information System (ACIS) maintained by the RCCs. Each drought index is interpolated to a grid, and grid points are aggregated to a spatial region of choice (e.g., county, hydrologic basin, drought management region, climatic division). The uniqueness of the DDIT is the ability for its users to select the measure of drought (e.g., Palmer indices, standardized precipitation indices), the time-frame (day, week, month), and spatial unit that matters most to them (Figure 8.1). Results are displayed on web pages using standard web-based scripting languages—HTML and PHP—and embedded scalable vector graphics (SVG). The interface provides users with a typical suite of geographic information system functions: they can navigate, change scale, highlight or conduct analysis on different layers, and change color and classification schemes. It also provides metadata to reveal the stations used to construct each choropleth map. The DDIT began as a collaborative effort between the University of South Carolina and the North Carolina and South Carolina state climatology offices. It was originally hosted on the South Carolina Department of Natural Resources (SCDNR) web server. Its evolution included partnerships with the Northeast RCC and the Climate Assessment for the Southwest (CLIMAS)—part of NOAA's RISA program.

Flexibility of this drought monitoring tool attracted attention outside the Carolinas, especially in the Southwest. The U.S. Southwest, a semi-arid region of high interannual and decadal precipitation variability, is prone to drought on time scales ranging from seasons to decades [4]. Since 2000, Arizona and New Mexico have experienced extensive drought impacts, including the loss of over 20% of their conifer forests from a combination of severe drought stress, increased insect outbreaks, and record-breaking fires [5]; loss of habitat for native freshwater fish (e.g., the Rio Grande silvery minnow); water restrictions in small cities, such as Santa Fe, NM, and Flagstaff, AZ; water shortages in vulnerable rural regions and on tribal lands [6]; and substantial livestock losses. The region's rapidly growing population makes it particularly sensitive to drought as it demands water from over-allocated rivers, such as the Colorado and Rio Grande. The rapidity with which drought impacts accumulated, as well as the persistence and severity of impacts, prompted the development of state drought plans (NM in 1999, AZ in 2004), the first-ever shortage criteria for Colorado River reservoir operations [7], and increased demand for drought information by county, state, federal, and private resource managers. Nevertheless, state funding for improving availability of drought information and putting this information into decision contexts germane to regional stakeholders has been lacking. The expansion to 18 eastern states followed from a similar interest in regional- and local-scale monitoring. Given the tool's use of ACIS data, it made sense to collaborate with the RCCs that maintain ACIS and serve as well-established providers of climate data services. The DDIT expansion grew out of both needs expressed by stakeholders for easier access to data and information to represent drought status at scales more

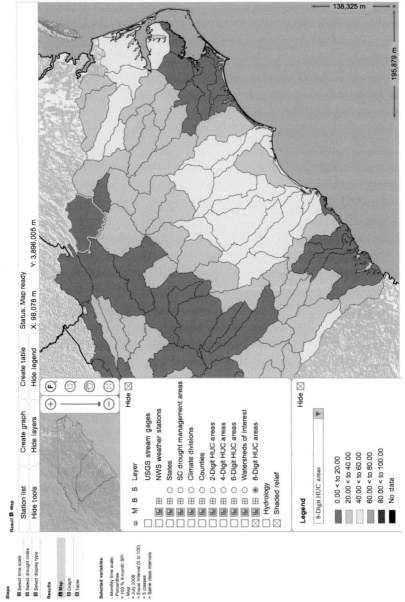

Figure 8.1 Sample output from the DDIT illustrating map display options and metadata listing user-selected parameters.

relevant to decision-making, and out of an interest in demonstrating the transferability and scalability of the tool as a pilot project for NIDIS regional drought early warning systems.

8.2.2 Technology transfer

Expansion of the DDIT revealed several software development issues that often are not considered in research applications. The literature addressing the software life cycle and the technology transfer process provides a framework for understanding the issues associated with collaborative software development and its implementation, diffusion, and adoption [8–11]. It addresses the inherent differences between research and operations, and of steps toward collaborative development that overcome these differences. The challenge is to have a research and development environment that encourages innovation and risk-taking versus an operations environment that requires stability and reliability [10–13]. Crossing the so-called "valley of death" between the development of decision-making tools and their successful dissemination involves overcoming technological, economic, and institutional challenges [11,14]. Technological limits may emerge from product software or hardware demands that an operational institution cannot support. Economic limits are often evident in either an inability to support the costs of transitioning from a research to an operations setting or the costs of maintaining tools at operational institutions. Institutional limitations involving the cultural practices of both research and operations groups and their broader missions also play a role [11].

The National Research Council (NRC) prescribed a pathway to overcome these challenges that starts with a proof of concept, follows with test-bed analysis and feedback, and culminates in the development of a research, development, and transition plan leading to full transfer into operations [13]. The pathway has been refined into three life cycle steps, each involving evaluation [14]. The first involves an examination of user needs, including information on spatial and temporal resolution, accuracy, data format, technological constraints, and user abilities [11,15]. The second step involves the development of a prototype to verify and validate results and gather feedback. Focus groups of end users can facilitate and strengthen this portion of the transitional pathway by demonstrating use, value, and validity of a product while providing feedback to researchers [13]. The second step should also include a formal development and transition plan that specifies the tool's resource needs, and a data management plan that includes data use, transition of raw data into information, metadata, distribution, compatibility, archiving, and user access [13]. Sound transition plans comprehensively describe each activity and require financial, technological, and human resources. They specify how each activity would be accomplished, state who is responsible for each task, and establish a schedule with measurable benchmarks [13]. The final step culminates with production and the eventual deployment of a new product. Four important questions arise in this final phase. Is a transition plan set up so users can efficiently employ the developed technology? Is there a plan set up to deal with the data involved in the project, and can the project accommodate increased use? Is there support for novice users who need training? Finally, and of greatest concern, can the developed product be sustained?

Beyond maintenance, success of a decision-support tool is often measured by adoption and user satisfaction [11,14]. Adoption requires diffusion, and diffusion is influenced by communication about a product's advantage, compatibility, complexity, trialability, and observability [12]. New tools must demonstrate improved cost, performance, or functionality over existing choices. They must mesh with the needs and organizational structures of a user community. The perceived difficulty in learning, understanding, and using a product influences a product's adoption, as does the ability to try a product before committing to it completely. Finally, the product must demonstrate visible and favorable results that are observed and discussed within a user community. Attention to the factors in Rogers' [12] framework should improve the chances of achieving operational success.

8.2.3   Collaborative development

Developing effective decision-support tools requires collaborative efforts among team members with various skills, working in tandem and across geographic space [9,16,17]. The development and implementation of the DDIT included climate scientists, climate impacts specialists, decision-support designers, programmers, and systems analysts located at many institutions. The group developed high levels of trust and communicated ideas and suggestions frequently and effectively. They often functioned in tandem, particularly members of the technology development team who contributed actively to software code development. Following recommended practices [9,18], they used remote servers that hosted the project and could be accessed from any location; this facilitated collaborative development within virtual teams. The technological complexity of research products like the DDIT requires organizational management and consistency in procedures, tools, software environments, naming conventions, and versioning [18]. Coding demands particular attention to detail to ensure compatibility and durability. Scalability of code offers another challenge affecting everything from functionality of a product to traffic during its operational phase. Here, adherence to a data-management plan can facilitate the creation of a scalable product—one that can accommodate growth [13]. Finally, product development needs to maintain the flexibility to operate within diverse computer environments and to keep pace with technological improvements [13].

8.3   User needs, technical considerations, and collaboration

Refining the DDIT involved feedback from users or potential users, and required technical changes to accommodate the long-term practices at the host institution. The adjustments were designed to make the tool as useful as possible to the decision-making community and also to make it more flexible and stable. Such transitions demanded collaborative effort between developers, the intended host of the tool, and potential users. Those involved in the DDIT transfer included climatologists, social scientists, and computer scientists from universities and state and federal agencies. Collectively, the project team represented both research and operational organizations.

8.3.1   User needs informing the process

The team solicited stakeholder feedback on DDIT usefulness and usability—working with the Arizona Department of Water Resources' Drought Program to identify drought information users representing a variety of decision contexts—to evaluate changes needed to make DDIT usable in the Southwest. The team built on CISA's prior [3] and ongoing work with stakeholders in the Southeast. This interaction was fostered by the South Carolina State Climatologist both in the development of the DDIT and the user groups that were involved in the FERC relicensing process and spin-offs that included drought management advisory groups interested in relevant drought intensity measures. This feedback was used to increase usability and improve the DDIT interface, by expanding DDIT source code, and facilitating shared programming and decision-tool implementation in a cooperating operational institution.

Team members directed the DDIT to two diverse focus groups of Arizona stakeholders to elicit feedback on the characteristics of the DDIT website. These characteristics included transparency of the procedures used to generate output (e.g., interpolation methods, metadata describing inputs, etc.), a need for additional map layers and information, translation of technical terminology, and a need for additional analytical tools. Participants represented cooperative extension workers, universities, federal and tribal natural resource management agencies, state environmental quality and water management agencies, and county and city water management and emergency preparedness agencies. Focus group participants completed a survey to provide the team with background information about their management and decision contexts, use of drought and climate information in decision-making, and professional training. All features of the DDIT were then demonstrated, followed by a hands-on exercise during which the focus group participants examined the website in pairs. Participants reconvened for a guided discussion. The research team recorded all stakeholder feedback.

Participants' chief drought-related concerns included ecosystem health, streamflow, surface water and groundwater supplies, water quality, and wildlife management. Participants found the DDIT to be a powerful and relatively easy to understand decision-making tool. All mentioned that it would help them in their work, specifically for impact analyses (e.g., for endangered species), public service announcements, and water resource decisions, as well as to justify management actions (such as policy or drought declarations). Participants expressed enthusiasm for the ability to custom blend drought indices, one of the most sophisticated capabilities of the DDIT. They noted that the DDIT may be overwhelming for some users, such as small water providers. Therefore, they recommended that the development team offer default options for nonexpert users. Participants made many recommendations for improvements to the DDIT and to enhance its usability in the Southwest. We categorized these improvements as interface enhancements, data and data management enhancements, and ancillary changes to improve comprehension and overall usability. Improvements most relevant to NIDIS and technology or knowledge transfer are indicated in Table 8.1.

Each of the suggested changes was then evaluated on the basis of technical challenges, resource needs, capability to match with available technical team skill sets, and value to stakeholders (i.e., is the requested refinement a need that is fundamental to adoption of the tool, or is it an incremental enhancement of a feature that currently works well but could be slightly modified?). The team then prioritized each action, based on consensus regarding the aforementioned criteria, and developed a timetable for implementing changes.

8.3.2 Technical considerations

Transitioning the DDIT to the Regional Climate Centers' ACIS-WS version 2 required two major adjustments to the original tool—ingesting gridded

Table 8.1 Suggestions for DDIT revisions from focus groups.

Interface
- allow comparisons of multiple maps or graphs on one screen
- retain user preferences between sessions
- allow user-generated labels and notes
- display river and mountain names with "mouse rollover"

Data
- include additional parameters: snow, groundwater
- add new spatial boundaries: physiographic, ecosystem, soil type regions, state trust lands, township and range, tribal land holdings, federal lands, and resource conservation districts
- allow importation and overlay of seasonal climate and streamflow forecasts

Ancillary
- display measurement units explicitly
- allow distance measurements across the map interface
- create a step-by-step tutorial or guidebook
- create a glossary and translation of technical terminology
- allow functionality for downloading GIS shapefiles or for exporting DDIT maps into desktop GIS software

input and interfacing with web services access routines that are part of ACIS. These requirements reflect the fundamental operational climate service needs for accommodating multiple applications and efficient generation, management, and dissemination of data. The DDIT's spatial features match most of those already used by gridded ACIS data sets, although a few specific spatial units needed to be added (e.g., specific watersheds and drought management areas unique to each state). While metadata for spatial features and climatic data have always been stored in the ACIS server, and retrieved upon users' request for the existing DDIT, the new version of the DDIT retrieves them more efficiently by using web services.

In some cases, the structure of the DDIT conflicted with ACIS-WS version 2. For example, the DDIT was originally designed to work from station data and store pregridded drought indices. Migration to ACIS-WS version 2 required gridded temperature and precipitation data to produce on-the-fly calculations of drought indices through web services. This evolution required investigation of not only how the choice of data sets and spatial data handling influences the output of drought indices, but also how these different paths affect DDIT performance. Such testing required time and altered data formats and analysis methods. The transition to web services was a major modification. Server-side SVG rendering and generation of multiple output formats including Portable Network Graphics (PNG) images, Keyhole Markup Language (KML) files, stand-alone SVGs, and embedded SVGs were also required. Each of these changes required significant time and demonstrated the importance of anticipating data structures and protocols of an operational application. The original DDIT's open-source and cross-platform structure facilitated code transfer to a virtual server. We successfully implemented functionality of the DDIT by exporting SVG maps into ESRI shapefiles, Google Map's KML format, and JPEG image files.

8.3.3 Team collaboration

The project team met monthly by teleconference, with several intensive work periods, to address key technical issues, identify technical priorities, and assign programming tasks. The team was divided into technical and social science working groups, and employed a social science team member to document the technology and knowledge transfer processes. The technical team developed climate, hydrology, and spatial information databases; augmented the existing DDIT website; and incorporated initial usability feedback from end users. It then worked with the Northeast Regional Climate Center to implement the DDIT on their server. The social science working group developed written survey instruments and focus group protocols for garnering feedback from stakeholders.

The technical working group used a scaling approach to broaden the Carolinas DDIT to accommodate users in the Southwest. Scaling involves refining the software code through an open-source process to accommodate (1) multiple programmers, (2) different geographies, and (3) regionally customizable features. Scaling can be contrasted with an "entire code transfer," whereby entities in one part of the country can maintain a separate copy of the code from entities in other parts of the country. The scaling approach makes it easier both to share corrections to errors ("bugs") and updates, and to add new functionality. Its foundation was to place the code on one server, and to allow programmers to access the code from several locations. Such scaling engaged a team of researchers, product developers, and operational entities in ongoing knowledge exchange and learning. It also fostered a sense of co-ownership among partners.

The Northeast Regional Climate Center agreed to collaborate with us and serve as a repository for the DDIT source code, provided our process conformed to its information technology protocols. Its protocols included a version control system that allowed multiple developers to simultaneously modify the same code. Each developed version of the DDIT is stored in the repository. Using this system and following its stringent industry-standard documentation guidelines allowed multiple developers to make modifications to the DDIT in a way that allows for tracking, alteration, and synthesis of software changes. It also avoided any interruption to the RCC's information technology system and eliminated maintenance costs and security risks. By using this approach, the team also acknowledged the important role of product maintenance, which is perhaps the most critical and expensive part of the product life cycle [19]. Of course, the scaling approach requires management oversight to track developments, avoid conflict, and reduce the likelihood of duplicate efforts. This management was provided by a technical team scientist–information technology liaison, who assigned rules for code development and refinements. For example, considering the software code as a tree, all changes must be made in branches, thoroughly tested, and only then, merged into the trunk. In addition, should the trunk break, someone should to be able to isolate and back out the offending work to restore the stability of the code base, so that development can continue. Mutually agreed-upon development rules are the cornerstone of the scaling process. In our estimation, this approach, while slow on the front end, results in greater efficiency and lower likelihood of duplicated effort and of error than the "entire code transfer" approach.

8.4 Lessons learned in the DDIT transfer process

Each part of the DDIT transfer process—user input, technical considerations, and team collaboration—offered valuable lessons that extend beyond our specific example. The crux of the transfer process was amending the software code structure to make the product robust for operational use. In the following discussion, we enumerate aspects necessary for this move and generalize some technical aspects of service innovation, as well as the learning aspects that led to enhanced knowledge exchange between researchers, product developers, and operational entities.

Data. Data issues can be quite substantial. Data-quality assurance, data interoperability, database maintenance, and provision of sufficient metadata are both essential and resource-intensive [20]. We discovered that users sometimes attribute greater accuracy to high spatial resolution data sets, regardless of problems associated with data interpolation to finer scales. Consequently, decision-support tools must issue disclaimers regarding data accuracy, homogeneity, missing value estimation, and the scale at which interpolated values are credible. In this regard, explicit metadata are essential.

Similarly, in contrast to research applications that typically rely on static data sets, operational tools depend on constantly changing real-time data and on the subsequent quality control changes to the data. This dynamic had to be factored in while designing the operational tool. When limited to the Carolinas, it was feasible to use the ACIS data stream to generate new sets of gridded data layers to account for new and modified data. However, as the geographic and temporal scope of the DDIT expanded, and as we added new input data streams, this approach became untenable. We amended the development plan to facilitate on-the-fly calculation of the gridded data, based on the most complete and highest quality data.

Efficiency. Code efficiency, and hence decision-support tool performance, must be a high priority in developing operational decision-support software. Performance speed factors prominently in the perceived usability of the tool, as users expect near instantaneous response, particularly from tools accessed via a web platform. Some questions on efficiency were heightened when considering the expansion of the DDIT beyond the Carolinas. In particular, issues of data storage versus on-the-fly calculations were complicated by institutional data-handling practices conflicting with precalculated drought indices and gridded data layers.

Flexibility. Decision-support tools also need a flexible design to adapt to new operating systems and changing paradigms in commercial software tools. Rather than constraining the DDIT to work, solely, with a particular database or version of software, we used nonproprietary software products. Developers must anticipate a continued need to update and maintain code. This requires sustained support to migrate the code to new technologies. In the experimental software development environment most frequently inhabited by RISA teams, this consideration is inconsequential. However, to become a viable operational tool, the financial, technical, and human resources costs of maintaining and updating the code must be considered.

Technology liaison. Operational agencies may not have sufficient capacity to anticipate future technological needs—such as new industry standards and technologies—due to barriers imposed by organizational culture [21] or by lack of investment in and attention to emerging information technology practices [22]. To build this capacity, operational agencies must acquire expertise in computer sciences, industry-standard practices, and personnel who can serve as a liaison between scientists and information technology specialists. In our experience, the decision-support tool development process can drive changes within operational entities in the technologies that they use and in the operational practices that they need to adopt, which often lag industry standards. In addition to subject matter experts, research teams require a skilled computer science expert to bridge the gap between research-experimental design and operational tool implementation for sustained success in the transition of research to operations. Someone in this capacity must actively monitor and assess technology market trends and develop plans to adapt to changing paradigms in commercial software development. This type of team member can help researchers and operations practitioners link needs and capabilities, and can foster collaboration among team members and partners [11].

The operational home. Based on our experience gained through extending and scaling the DDIT, we found that operational organizations need the following personnel to handle the transfer from research to operations: a full-time system administrator, a professional software developer, a subject matter professional, and a commitment to maintain data and software. In addition, the organization will need to have sufficient hardware and budget to maintain and sustain the decision-support tool. For example, the SCDNR, the project's original operational partner, did not have the institutional mandate and mission to accommodate the multistate DDIT expansion required by this project. Institutional stability is also required [21]. University programs and departments often lack long-term staff stability, stable funding sources, and sufficient system administration and information technology resources within individual departments. They are not always ideal operational bases for long-term, decision-support products. Further, to facilitate the transfer to a qualified operational home, researchers and developers must be able to demonstrate to the operational home that they can work collaboratively in an information technology context, meet the home organization's security requirements, and not damage the home organization's mission. Frequent and in-depth conversations between researchers-developers and operational staff must address technical capabilities, operational practices, information technology capacity, budget, and long-term financial commitments of operational agencies developing experimental decision-support tools. Operational entities must explicitly state their standards, constraints, and willingness to commit to upscaling and expanding decision-support tools outside their current geographic region and private industry and public sector users as well as support research and experimental modes. Support by NOAA, or other federal entities, can leverage investment in existing institutions such as the Regional Climate Centers to serve as operational homes for newly developed software tools [23].

Documentation. We found that the mutual benefits, as well as costs, to the successful research-to-operations public sector technology transfer can be substantial. For example, the documentation necessary for multiparty software development and transfer forces developers to make explicit every aspect of their software code, databases, and development process. Documentation requires a large investment in staff time and meetings, but competent documentation allows others to more easily adopt and use the tool, make modifications, and upgrade it as necessary. In our opinion, collaborative development, using the shared code model, in which the experimental product developers and the operations team work together, provides greater benefits, despite high initial costs, than using off-the-shelf software and/or hiring a consultant. In the former case, all parties gain a better understanding of the research to operations development process. Improved communication and dialogue about user needs, technology innovations, and other topics foster continued tool development by expanding capabilities and providing new functionality.

Negotiating conflict. Both researcher and operations teams need to be prepared for conflict. For example, use-inspired research may lead to product interface and navigation demands that cannot be accommodated due to technical standards or institutional constraints. Similarly, operational entities may benefit from these demands to develop new capabilities, such as using open-source code and version control systems, to successfully adopt a decision-support tool.

Maintenance, validation, and evaluation. Stakeholders make a substantial investment in learning to use a decision-support tool. To honor that commitment, researchers, developers, and operational teams must refine and update tools while preserving continuity of service. This challenges operational entities that lack a wide range of skills. These challenges need to be adequately accounted for within budgets and in the project zeitgeist. Tasks such as analyzing software performance statistics, revising the code to remove bottlenecks, and developing capacity to serve greater numbers of users are all integral parts of the process. Developing sophisticated graphic designs and performing careful analyses of the user interface require skills and resources that cannot be adequately supported by small, rapid turnaround grant funding. The DDIT project described herein required six different technical specialists—graphic designer, program developer, GIS specialist, technologist, database specialist, and system administrator—in addition to subject area and data experts. A staff of this size, which is presumably needed for other technology transfer projects, requires commensurate resources and an on-going support model.

Scaling and extending. Scaling a tool refers to adapting software and hardware to accommodate increases in the daily number of visitors to the site. Similarly, scaling the application refers to expanding the situations in which the tool will work, such as changing data sources or adding data for additional regions. Extending the tool refers to incorporating new functions or features in the tool, regardless of changes in the number of users. In this project, the DDIT was both scaled up—to accommodate users in other regions—and extended—to incorporate new functions and enhance usability. Scalability and extendibility are facilitated by an object-oriented paradigm, modular code development, open source coding, and a comprehensive initial design that facilitates these kinds of extensions. Approaches that foster extendibility and scalability—in short the ability of a software to accommodate modular change—demand up-front investments in human resources that will ultimately avoid a more expensive product development investment that may result in stalling the operations gateway or requiring remedial effort to correct earlier flaws. Within atmospheric sciences, the global climate modeling community has set a successful precedent for the kind of programming practices that foster extendibility and scalability.

Environment for virtual collaboration. Open-source programming with contributors from multiple institutions runs the risk of upsetting the operational system of the institution hosting a transferred experimental product [8]. Remote host servers, which provide a virtual collaborative environment for programming, allow the product development team to examine and repair compatibility issues and other potential sources of error without the risk of causing damage or interruption to the sponsoring institution's information technology system. The remote host server/remote institution system also provides a reduced-cost mechanism for temporary use of the substantial hardware infrastructure needed for developing decision-support tools in a secure environment. The DDIT project used a remote server to evaluate upscaling issues, such as alterations to the code, to remove bottlenecks in processing time when many users access DDIT simultaneously.

Project planning, documentation, and standards. A double-edged aspect of the remote institution process is that it requires the development-operations team to impose strict rules of engagement (standards) to ensure that tasks, errors, and revisions are well tracked and well documented [9,10]. One downside to imposing the discipline necessary to employ these methods is that the learning and internal deadlines may inhibit use of short-term student labor or agency-based atmospheric scientists who do not have formal credentials in computer sciences. The silver lining is that the well-defined processes form an excellent foundation for collaborative software development.

Mutual understanding of the software development life cycle. In scientist-developer-operations decision-support tool development and transfer, everyone involved in the transfer must have a clear understanding of the industry standard for the software development life cycle from the beginning of the project [10]. This gives the entire team a clear set of guidelines and a clear sense of milestones in the process, which will avoid ad hoc development approaches. For the DDIT project, life cycle learning involved multiple intensive meetings and a systematic assessment to bring scientists and IT personnel to a common understanding of what tasks were needed to implement user-inspired recommendations as well as of the order in which the tasks needed to be accomplished. Adhering to a strict process can generate frustration, especially on the part of operations staff used to moving quickly when adopting new software. Adopting and implementing experimental software requires learning on both sides, setting expectations, and adjusting to an often unequal relationship between subject matter scientists and computer scientists, each of whom must respect the other's expertise [9].

In summary, our review of literature related to technology transfer and our specific experience in expanding and transferring a drought-monitoring tool suggest that several metrics help to assess the required collaborative process. First, a formal software development process that includes effective communication among scientists, developers, and operations personnel is essential. Development and implementation require interdisciplinary project members that understand each other. Second, the development process must assure high data quality standards and portability to a new host, and include performance tuning and a process for handling unexpected issues (e.g., data problems). Third, the process should include user feedback to facilitate usability and comprehension of the tools developed.

In addition, technology transfer must adhere to software development standards that include tools for efficient code development and version control, a task trading system that facilitates collaborative development, and a code review process. Implementing standards is essential and requires broad-scale agreement among collaborators within a software development document, a coding standards document, system-specific software tools, and an integrated development environment. The collaborative development process for transfers of experimental tools to operations *is part of the science translation activity*. Moreover, it provides an opportunity to train a new generation of researchers who understand computer science standards, scientific and experimental software innovation, and response to stakeholder needs.

References

1 109th Congress (2006) *Public Law 109-430, National Integrated Drought Information System Act of 2006.* URL http://www.gpo.gov/fdsys/pkg/PLAW-109publ430/html/PLAW-109publ430 .htm [accessed on 22 July 2013].

2 NIDIS Implementation Team (2007) *The National Integrated Drought Information System implementation plan (NIDIS): A Pathway for National Resilience*, 29 pp. NIDIS Program Office, Boulder, CO. http://drought.gov.

3 Carbone, G.J., Rhee, J., Mizzell, H.P. *et al.* (2008) A regional-scale drought monitoring tool for the Carolinas. *Bulletin of American Meteorological Society*, **89** (**1**), 20–28.

4 Hoerling, M.P., Dettinger, M., Wolter, K. *et al.* (2013) Present weather and climate: evolving conditions. In: Garfin, G., Jardine, A., Merideth, R., *et al.* (eds), *Assessment of Climate Change in the Southwest United States: A Report Prepared for the U.S. National Climate Assessment.* Island Press, Washington, DC, pp. 74–97.

5 Williams, A.P., Allen, C.D., Millar, C.I. *et al.* (2010) Forest responses to increasing aridity and warmth in the southwestern United States. *Proceedings of the National Academy of Sciences,* **107**, 21289–21294.

6 Ferguson, D.B., Alvord, C., Redsteer, M.H. *et al.* (2011) *Drought Preparedness for Tribes in the Four Corners Region Workshop Report.* Climate Assessment for the Southwest, Tucson, AZ, 36 pp. URL http://www.climas.arizona.edu/files/climas/pubs/four-corners-drought-preparedness-2011.pdf [accessed on 22 July 2013].

7 U.S. Department of Interior (2007) *Record of Decision. Colorado River Interim Guidelines for Lower Basin Shortages and Coordinated Operations for Lake Powell and Lake Mead.* URL http://www.usbr .gov/lc/region/programs/strategies/RecordofDecision.pdf [accessed on 13 July 2008].

8 Constantine LL. (2001) *Beyond Chaos: The Expert Edge in Managing Software Development,* 392 pp. Addison-Wesley, Boston.

9 Evans I. (2004) *Achieving Software Quality Through Teamwork,* 296 pp. Artech House, Boston.

10 Stellman A, Greene J. (2006) *Applied Software Project Management,* 308 pp O'Reilly, Sebastopol, CA.

11 Brooks, M.S. (2013) Accelerating innovation in climate services: the 3 E's for climate service providers. *Bulletin of the American Meteorological Society,* **94**, 807–819.

12 Rogers, E.M. (1983) *Diffusion of Innovations.* The Free Press, New York.

13 National Research Council (2003) *Satellite Observations of the Earth's Environment: Accelerating the Transition of Research to Operations.* National Academies Press, Washington, DC. ISBN-10: 0-309-08749-X.

14 Gilruth, P.T., Kalluri, S., Robinson, J.W. *et al.* (2006) Measuring performance: moving NASA Earth science products into the mainstream user community. *Space Policy,* **22** (**3**), 165–175.

15 Jacobs, K.L., Garfin, G.M. & Lenart, M. (2005) More than just talk: connecting science and decisionmaking. *Environment,* **47** (**9**), 6–22.

16 Shim, J.P., Warkentin, M., Courtney, J.F. *et al.* (2002) Past, present, and future of decision support technology. *Decision Support Systems,* **33** (**2**), 111–126.

17 Joshi, K., Sarker, D.S. & Sarker, S. (2007) Knowledge transfer within information systems development teams: examining the role of knowledge source attributes. *Decision Support Systems,* **43**, 322–335.

18 Tuohey, W.G. (2002) Benefits and effective application of software engineering standards. *Software Quality Journal,* **10**, 47–68.

19 Zelkowitz, M.V., Shaw, A.C. & Gannon, J.D. (1979) *Principles of Software Engineering and Design.* Prentice-Hall, Englewood Cliffs, NJ.

20 National Research Council (2009) *Restructuring Federal Climate Research to Meet the Challenges of Climate Change,* 178 pp. National Academies Press, Washington, DC.

21 National Research Council (2009) *Informing Decisions in a Changing Climate.* Panel on Strategies and Methods for Climate-Related Decision Support. Committee on the Human Dimensions of Global Change, Division of Behavioral and Social Sciences and Education, 188 pp. National Academies Press, Washington, DC.

22 National Research Council (2010) *Informing an Effective Response to Climate Change.* America's Climate Choices: Panel on Informing Effective Decisions and Actions Related to Climate Change, 325 pp. National Academies Press, Washington, DC.

23 DeGaetano, A.T., Brown, T.J., Hilberg, S.D. *et al.* (2010) Toward regional climate services: the role of NOAA's regional climate centers. *Bulletin of American Meteorological Society* **91**, 1633–1644. 10.1175/2010BAMS2936.1.

CHAPTER 9

Managing the 2011 drought: a climate services partnership

Mark Shafer[1], David Brown[2] and Chad McNutt[3]

[1] Southern Climate Impacts Planning Program, University of Oklahoma, 120 David L. Boren Blvd., Suite 2900, Norman, OK 73072, USA

[2] National Centers for Environmental Information, National Oceanic and Atmospheric Administration, Fort Worth, TX 76102, USA

[3] National Integrated Drought Information System Program Office, University Corporation for Atmospheric Research, Boulder, CO 80307-3000, USA

9.1 Introduction

Wilted crops, dry ponds, forests ablaze. These are the images of drought. However, a drought has many more subtle, pernicious, and long-lasting impacts too. The economies of rural communities are devastated by the lack of spending at local businesses, restaurants, and hotels when there is nothing to harvest [1]. Emotional and psychological effects take their toll on individuals faced with the possibility of losing their livelihoods. Social disruption and dislocation are possible in extreme cases as people abandon or sell land and relocate in search of better opportunities.

Understanding these long-lasting effects and helping people manage during drought has been a growing area of research and services. The National Integrated Drought Information System (NIDIS), established by Congressional mandate [2] to coordinate drought-related information and services, and the National Oceanic and Atmospheric Administration (NOAA)'s Coping With Drought Initiative have focused resources on studying information flow between operational forecasters, the monitoring community, and those affected by drought. NOAA's Regional Integrated Sciences and Assessments (RISA) is a part of this network, along with NOAA's Regional Climate Services Directors (RCSDs) and the National Drought Mitigation Center (NDMC).

The 2011 Southern Plains drought caused an estimated $12 billion in damages [3], primarily in Texas, Oklahoma, and New Mexico (Figure 9.1). To improve information flowing from NOAA and other sources and to

Climate in Context: Science and Society Partnering for Adaptation, First Edition.
Edited by Adam S. Parris, Gregg M. Garfin, Kirstin Dow, Ryan Meyer, and Sarah L. Close.
© 2016 John Wiley & Sons, Ltd. Published 2016 by John Wiley & Sons, Ltd.

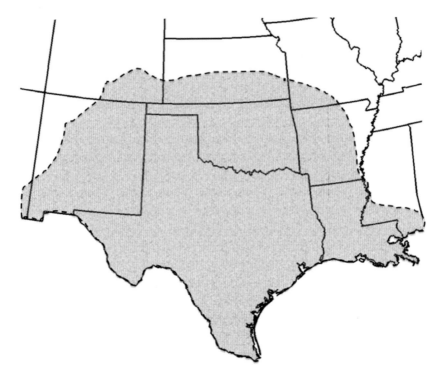

Figure 9.1 Approximate area affected by severe to exceptional drought during 2011 in the Southern Plains. Primary area of services during the 2011 drought focused on Texas, Oklahoma, and Eastern New Mexico. Additional services were provided to surrounding states.

better understand the challenges of managing social, ecologic, and economic systems in the face of these extremes, two RISA teams, the Southern Climate Impacts Planning Program (SCIPP) and Climate Assessment for the Southwest (CLIMAS), the NOAA RCSD and NIDIS worked together in a concerted effort to connect these two challenges. Activities included six regional drought forums, more than a dozen drought webinars, weekly status and outlook briefings, a region-wide planning workshop, and an effort to improve drought monitoring. This chapter will examine the unique aspects of each component, why all components were necessary, how they informed each other, and what providers and scientists learned from these interactions.

9.2   Evolution and impacts of the drought

The drought that developed in the Southern Plains in 2011 and then extended into the Midwest in 2012 is among the most severe in recent history. The drought began in the Fall of 2010 through the convergence of several large-scale oceanic and related atmospheric circulation patterns. A similar alignment occurred in the 1950s, which is the drought of record for the Southern Plains [4,5]. The dry circulation pattern was aided by La Niña, which tends to favor dry winters and springs in the Southern Plains. A strong La Niña event developed during the Fall of 2010. Without the typical late winter and spring rainfall, soil moisture went into summer already depleted. Instead of evaporating water from the soil, the sun's energy heated the land and atmosphere, resulting in some of the hottest and driest months since 1895 [6].

Over the next 3 years, drought continued in parts of the region, although generally not as extreme as experienced in 2011 (Figure 9.2). Following a relatively wet winter in 2011, during the ebb-and-flow of drought in central and eastern Texas and Oklahoma, western parts of both states and eastern New Mexico remained mired in extreme and exceptional drought conditions. This area remained nearly continuously in severe drought conditions since the drought first developed in Spring 2011, until it ended in early 2015, leaving multiple years of rainfall deficits and conditions in some places resembling those of the 1930s Dust Bowl.

The most immediate impacts during 2011 were a nearly complete failure of the winter wheat and summer crops, large-scale cattle sales and relocation, transport of hay on an unprecedented scale, extreme wildfires, and ecological impacts. Agriculture-based communities struggled as cattle was sold or relocated, farms failed to produce crops, and local businesses were unable to generate revenue. Municipalities faced critical water shortages. The natural environment suffered as hundreds of millions of trees died, wildfires raged, and blue-green algae bloomed in area lakes. Impacts of the drought on water resources, agriculture, ecology, and the economy lingered into 2015.

Each of these sectors needed information on the drought's status and outlook, but many needed information at different spatial and temporal scales. Agriculture needed estimates of likely precipitation for crop production strategies. Ranchers needed to know when pastures would recover for restocking decisions. Water resource managers needed to make investments in alternative water sources, with the duration of the drought heavily impacting those investment decisions. Public utilities needed to know if deep soils were recovering and whether to expect continued infrastructure damage in subsequent summers. Forestry needed estimates of how dry potential fuel sources were for decisions on fire prevention staffing and predeploying resources. Above all, it seemed that each of these sectors turned to different sources of information, just to make the picture more complex.

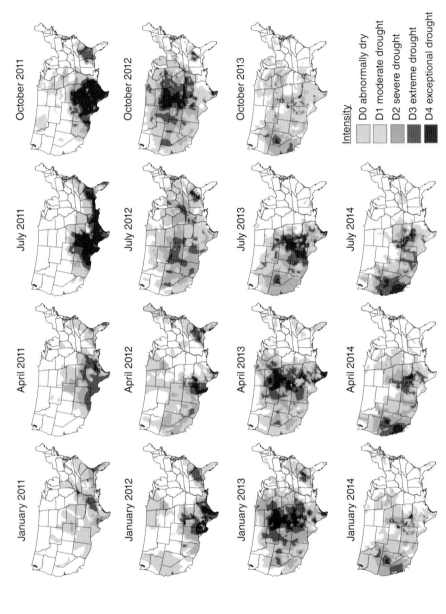

Figure 9.2 The progression of the Southern Plains drought as captured by the U.S. Drought Monitor. The most severely affected areas are indicated on the maps by the darker shades, according to the U.S. Drought Monitor scale. Source: U.S. Drought Monitor.

9.3 Existing climate services capabilities and building a drought community

Although the impacts could not be avoided, there were opportunities for RISA and its partners to provide information on the status and expected evolution of the drought thus helping communities and businesses to cope with the effects. The most important element was conversation. Creating forums for people to discuss the drought and its impacts and to interact with forecasters and scientists to talk about its evolution and outlook could at least open communication channels between scientists and practitioners.

The efforts built upon a drought knowledge community [7], composed of climate services providers and operational meteorologists, who had witnessed multiple previous drought events, particularly an 18-month extreme drought that occurred in 2005–2006. Many of these members also had participated in the U.S. Drought Monitor. Although there had been no previous concerted efforts on the scale that was undertaken during the 2011 drought, the primary participants had become familiar with each other and their respective expertise and maintained communication through interactions at meetings, discussion groups, and personal interaction. Drought knowledge community partners include:

State climatologists—Most states have a state climate office that provides data, analysis, and decision support for a range of stakeholders. State climate offices are able to conduct targeted research and compile a synthesis of existing research and data to respond to specific requests.

Regional Climate Centers—Regional Climate Centers manage large data sets and produce standardized products detailing historical climate variability, trends and extremes along with derived data sets such as various drought indices. Texas and Oklahoma are within the jurisdiction of the Southern Regional Climate Center and New Mexico is served by the Western Regional Climate Center.

NOAA RISA teams—RISA teams conduct research and analysis to support decision-making needs of specific stakeholders and to better understand those decision processes to inform partners on how to improve data sets and products. The 2011 drought region was primarily in a region serviced by SCIPP (Texas and Oklahoma) and CLIMAS (New Mexico).

NOAA National Weather Service, Climate Services Program Manager (CSPM)—The National Weather Service (NWS) has a CSPM located at all six regional offices. CSPMs are charged with initiating and coordinating regional partnerships and climate services activities. The CSPM for the Southern Plains is located in Fort Worth, Texas, at the NWS Southern Region Headquarters.

NOAA Regional Climate Services Director (RCSD)—NOAA established (RCSDs) to improve coordination among NOAA agencies and other partners. Texas, Oklahoma, and New Mexico are served by the Southern RCSD, based in Fort Worth, Texas.

National Integrated Drought Information System (NIDIS)—NIDIS operates a drought portal to synthesize relevant drought indices and planning information and conducts regional drought early warning systems that are designed to examine information needs and decision processes that can then be applied to serving products and information for other areas.

National Drought Mitigation Center (NDMC)—NDMC was established at the University of Nebraska-Lincoln in 1995 to help people and institutions develop and implement measures to reduce societal vulnerability to drought, emphasizing preparedness and risk management rather than crisis management. NDMC has expertise in climate-based monitoring, Geographic Information Systems and analysis, planning, and social science.

Together, these entities provided a range of services, from monitoring drought conditions to data analysis to planning and preparedness. To some extent, however, the effort lacked coordination. Stakeholders were not necessarily aware of some of these entities or the full range of services that were available through them. The 2011 drought brought all these partners together in direct connection with stakeholders through forums, webinars, impact reporting, and other activities.

9.4   Engaging stakeholders: multiple methods

The climate services partners had four goals in forming their services strategies:

1 to improve the flow of relevant information to affected individuals, communities, and businesses;
2 to learn what information was needed by those stakeholders;
3 to test replication of services from other locations; and
4 to assess the social, economic and ecological impacts of drought.

Multiple engagement efforts were launched with decision-makers from regional, state, and local arenas in a conversation about drought. These included regional forums, state drought planning, webinars, and engaging stakeholders in impact reporting. The net effect of these efforts was that the interaction between these arenas and between the academic and practitioner communities increased substantially. Indeed, many decision-makers participated in more than one of these efforts, such as state drought planners attending the regional forums or local Farm Service Agency representatives participating in the drought webinars and impact reporting. Direct communication between the two communities facilitated use of information, as opposed to one-way posting of information on websites or media advisories.

Participants were recruited through existing mailing lists by each of the partners and through contacting state organizations representing stakeholders, such as associations of conservation districts, water management districts, and emergency managers. The combined efforts of forums and webinars have reached more than 425 people, not including 49 who were part of various climate services provider institutions. Of these, 222 participated in at least one webinar, 99 of them in three or more. At least 164 people participated in a drought forum, several attending more than one even though the meetings were held at places separated by hundreds of miles. What is especially encouraging for building a drought community is the cross-fertilization between activities. Twenty-seven of the attendees of at least one forum also participated in at least one webinar. As an example of the inter-relationships of these activities, following the forum held in Fort Worth, Texas, two of the panelists were recruited to be presenters on subsequent webinars.

9.4.1 State drought planning

Within the six-state region served by SCIPP, three states (Louisiana, Arkansas, and Mississippi) have no drought plans; one state (Tennessee) has a plan solely focused on water resources; and Oklahoma's plan is more than 15 years old (adopted in 1997). A drought plan usually describes how an institution will monitor, mitigate, and respond to drought. In many cases, the plan will be associated with a series of triggers that establish a response or management action based on the severity of the drought [8]. To encourage them to create or update plans, SCIPP hosted a drought planning workshop, convened in Memphis, Tennessee, in May 2011. Workshop planning had begun prior to the severity of the drought becoming apparent. Thus, it was fortuitous circumstances, to some extent, that enabled engagement with some drought managers in parts of SCIPP's region.

Workshop goals were to:

1 introduce participants to the Drought Monitor process and NIDIS;
2 briefly discuss the strengths and weaknesses of various monitoring tools;
3 provide examples of good structure of state drought plans, including monitoring, communication, impact reporting, and connections to local communities; and
4 give them ample time to work with "experts" in outlining elements of their own (future) state plans.

In preparation for the workshop, SCIPP identified current or potential leadership on drought issues within each state. These officials were invited to the regional workshop along with drought experts in their respective states, including the State Climatologists and representatives from the Southern Regional Climate Center. National experts from NDMC, NOAA, and NIDIS attended and presented on a range of topics from the Drought Monitor process to planning resources. Academic institutions and federal agencies both contributed to the content and agenda.

Emerging social science techniques and facilitation methods helped the stakeholders identify and organize current challenges and future opportunities in drought mitigation and planning [9]. Through participatory learning and engagement, the following themes emerged within the group. The workshop highlighted a need for more monitoring tools and predictions; more analysis and coordination between sectors and agencies, even in the best prepared states; integrating drought into state water and hazard plans; and revisiting who and what agencies were involved in the original plan; and any necessary changes.

Outcomes from the workshop laid a firm foundation for further collaboration through the peer learning process. Participants learned from each other. There was as much conversation among the state participants as there was between them and the "drought experts." For example, participants from Tennessee and Oklahoma discovered that requests for well permits could be an indicator of emerging drought problems. Most importantly, participants now know their counterparts working with drought issues in nearby states and sources of expertise in regional and national organizations. Several members participated in additional national meetings and are working to improve drought management even in places where a formal planning process is in its infancy. SCIPP remains engaged with these participants, helping them identify counterpart plans, conduct surveys of agencies, and collect drought impacts information.

9.4.2 Regional drought forums

The centerpiece of engagement activities was a series of drought forums convened by NIDIS and the NOAA RCSD. Six in-person forums and one virtual forum were convened between 2011 and Summer 2014: Austin, Texas (July 2011); Fort Worth, Texas (November 2011); Lubbock, Texas (April 2012); Abilene, Texas (November 2012); Goodwell, Oklahoma (March 2013); Wichita Falls, Texas (June 2014); and Albuquerque, New Mexico (June 2012; virtual meeting). A major emphasis of the forums was improving the coordination and provision of drought-related information to critically affected sectors and communities. The forums focused on providing assessments of present drought conditions and impacts, comparisons with past drought events, and outlooks for the next season and into the following year. Representatives from water resources, agriculture and livestock, forestry and wildfire management interests, and state and Federal agencies and offices were among those in attendance.

All of these forums used a similar format made up of two sessions. The first session was considered the "drought outlook" and included a discussion on the evolution of the drought, factors in creating the drought's severity, current conditions, and how the drought might evolve over the coming season. Presenters included NOAA climate scientists, state climatologists, regional climate centers, and the RISA teams among others. The second session varied according to issues important to the locale, but generally included a mix of practitioners and researchers as presenters or panelists.

Forum organizers observed that the drought outlook session appeared to be most beneficial to stakeholders, who appreciated the in-depth discussion of the seasonal outlooks, including the reasoning and uncertainties associated with those predictions. The afternoon sessions yielded specific, usable information for climate services providers and highlighted areas where further research and collaboration was needed. Some common themes that emerged from the discussions were the need to clearly relate current conditions to historical conditions; creating simple graphics summarizing the current situation; being proactive with local reporters; improving drought-related triggers and indicators; and more analysis of the impacts of evaporation on reservoirs and crops. Much of this was background information for the participants, but several indicated specific use of information in their decision making.

Most agreed that there is a wealth of hydrological and meteorological information available on the Internet, but guidance on product interpretation, especially for understanding the drivers and the drought forecasts, and systematic assessments of impacts are needed. It was noted that posting products and narratives understandable to the layperson would be an improvement. State Cooperative Extension Agents were identified as an example of a group that provides guidance and interpretation with their products and therefore were perceived as a useful data source regarding agricultural impacts and best practices.

Some forum attendees requested help obtaining data at the local or basin scale and guidance in preparing local data for analysis or presentation to the public. Additional discussions identified opportunities and practical means to improve delivery of information. All agreed that the forums had provided a positive step in that direction and that one of the first approaches should be to work with NIDIS to develop a best practices document for improving coordination among federal, state, and local agencies, as well as for improving how information is delivered for proactive decision making.

9.4.3 *Webinars*

Communication among agencies and affected sectors is a key to successful management. Toward this end, a bi-weekly webinar series was launched in Fall 2011. The series' goals included:

- To improve communication among agencies and organizations in the Southern Plains who were being affected by the historic drought;
- To provide information on available resources and assistance to help monitor and manage drought;
- To understand the impacts of drought in this region *from the perspective of those who were tasked with managing it*; and
- To document impacts that will help improve the weekly U.S. Drought Monitor assessment and our understanding of how drought impacts evolve and decay.

The webinars were developed as a way of maintaining momentum in the regional drought forums. The forums provided discussion of impacts and management, but they were comparatively expensive and difficult to hold on a frequent basis across such a large area as was affected by the drought. The webinars were a cost-effective approach that could reach a broad audience, although the benefit of direct contact was lost.

RISA teams borrowed from the approach NIDIS is using to develop drought early warning information systems across the United States [10]. NIDIS offered two models of webinars, one used in the Upper Colorado River Basin that was geared toward water management professionals and one used in the Apalachicola–Chattahoochee–Flint (ACF) River Basin in the Southeast United States geared toward more general audiences. Given the regional goals of informing a broad discussion of the drought among multiple sectors, the ACF format proved to be a better model for the Southern Plains region.

Each webinar was a pared down version of the drought forum agenda, including an overview of the current drought assessment and outlook (led by NDMC), summary of impacts across the region (led by State Climatologists), and a topic or resource, such as La Niña or wildfire conditions. Topics were a mix of technical and sector-specific information and presenters included both scientists and stakeholder organizations (Figure 9.3).

Information from each webinar, including the presentations and a two-page summary in pdf format was posted on the NIDIS Drought Portal

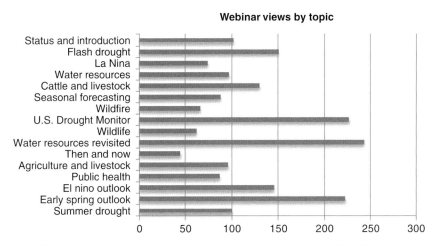

Figure 9.3 Number of views on YouTube for each webinar topic from earliest (September 2011) to most recent (August 2013).

and on the SCIPP website, Facebook page, and YouTube. Direct participation in the webinars averaged about 60–80 registrants on any individual topic through most of 2012. By late 2012, direct participation declined substantially, to fewer than 20 on most webinars; however, subsequent views on YouTube still averaged more than 100 per topic.

Each webinar took several days to plan, produce, and summarize, and there were difficulties in scheduling presenters. Frequent hour-long webinars also tested the patience of participants. To meet the need of frequent updates of drought conditions, a 5-minute recorded weekly briefing was implemented in 2013, increasing the frequency of the webinars. The briefings were posted directly to YouTube, alleviating the need for setting a scheduled time and coordinating presenters. The briefings were based off the slides used by the NDMC in the briefing part of the full webinars, which further traces its origins back to the format of the drought forums.

The weekly briefings drew about 70 views per week. This maintained interest and connections to our stakeholders in between the more detailed webinars, with fairly low overheads (about 30 minutes per week to produce). Keeping the connection was important because it allowed us to go to these stakeholders for further information and to recruit participation in other events, such as forums, studies, surveys, and other meetings.

9.4.4 *Impact reporting*
Impact reports are a critical factor in the U.S. Drought Monitor, which in turn is used as an eligibility criterion for several federal assistance programs

and a trigger for state drought response plans. Yet, most reports in the U.S. Drought Impact Reporter—a web-based mapping tool designed to compile and display impact information across the United States—are media stories or *ad hoc* reports [11]. The Drought Impact Reporter, and crowdsourcing in general [12], is an important approach for getting improved spatial representation of features, although potentially decreasing the quality of reports as opposed to using trained observers. SCIPP worked with state climate offices in Oklahoma and Arkansas to increase user reports during 2011–2012, as the drought was developing and changing rapidly (Figure 9.4). Since 2012, the long-term deficits have become the dominant drivers of drought in the region and monitoring tools are better equipped to capture those patterns. The addition of user reports added important details during the rapid evolution phase of the drought. These reports were coordinated with the NDMC and integrated with the archive in the Drought Impact Reporter (Figure 9.5).

A prime example of how impact information can be used occurred in Arkansas. Arkansas was affected by extreme drought conditions during fall and winter of 2010–2011. Drought conditions ended abruptly in April and May as excessive rainfall produced the flood-of-record in much of the state. However, by June, rainfall essentially shut off and temperatures soared. Moisture that accumulated from the April–May rainfall was quickly depleted from the soils and water levels in reservoirs began to fall. Because

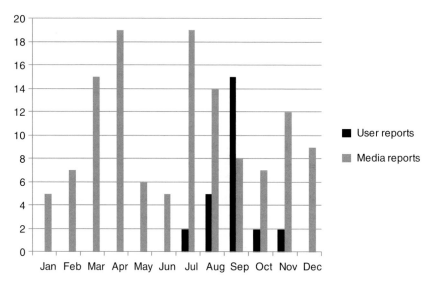

Figure 9.4 Number of reports in the Drought Impact Reporter for Oklahoma submitted during 2011 that were originated from media coverage (gray) and direct user submissions (black). Media reports represented here include only Oklahoma media; it does not include national coverage of events in Oklahoma.

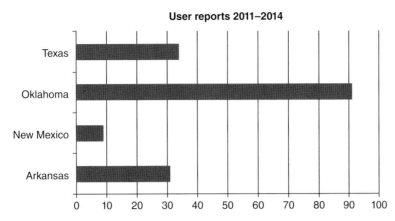

Figure 9.5 Number of user-submitted reports in the Drought Impact Reporter for 2011–2014 for the core drought area plus Arkansas. Climate Services partners initiated efforts to increase the number of user reports in Oklahoma and Arkansas during late 2011 and 2012.

of the excessive rainfall during the spring, many of the indices used in the Drought Monitor that are based upon precipitation departures over various time periods did not show severe drought conditions. With many indices pointing toward normal conditions and memories of the recent flooding, drought designations were gradually introduced into the state but remained moderate in most of the state (D1). On-the-ground reports painted a more dire picture, driven by the high evaporation rates that are not captured in many drought indices. Working with the Arkansas State Climatologist, SCIPP launched an effort to improve reports from district Farm Service Agency offices and irrigation districts. These reports pointed toward conditions similar to those seen further westward in the severe and extreme drought areas in Oklahoma, including nearly-dry farm ponds, lack of forage for cattle, and stunted crop development. These reports were provided to the Drought Monitor authors and subsequently helped to reassess conditions and align the depiction more closely to the reported impacts.

9.5   The important role of each climate services partner

Each partner brought some unique skills and expertise to the endeavor and all partners and all forms of interaction were essential and valued differently by each stakeholder. Partners included:

- *NOAA RISA teams*—SCIPP and CLIMAS used their stakeholder connections to promote the webinars and forums. Compared to operational Federal organizations, as university-based organizations, the NOAA RISA teams have more flexibility for testing innovative services in response to evolving stakeholder needs.

- *NOAA RCSD*—Provided coordination and organizational support that facilitated the development of the drought forums, provided connections to national media outlets, and leveraged to bring scientific expertise from various parts of NOAA to the meetings and webinars.
- *NIDIS*—NIDIS highlighted practices that were being tested in other areas of the United States, supported travel for stakeholders and presenters to the meetings, and conducted ongoing service assessment activities.
- *NOAA NWS CSPM*—The NWS, through the CSPM, provided drought monitoring expertise and ensured that NWS assets, such as the Weather Forecast Offices and the Climate Prediction Center, were fully used.
- *National Drought Mitigation Center (NDMC)*—The NDMC provided drought briefings for the webinars, highlighted relevant monitoring products and processes for webinar participants, and provided guidance in the regional planning workshop.
- *State Climatologists*—Nine State Climatologists (including surrounding states) participated in the webinars and forums, providing regular updates on the evolution of drought in their states. This gave a more local connection to an otherwise regional service.
- *Regional Climate Centers*—Regional Climatologists provided analysis of the drought evolution, access to drought indices tailored to the region, and presentations at the forums.

Each was able to contribute unique resources—expertise and financial or logistical support. By having all of these partners involved, it was possible to undertake the efforts with only a moderate impact on any one organization.

These efforts arose organically through ongoing conversations between the NOAA RCSD, the NIDIS Program Office, and SCIPP. The NOAA RCSD had recognized authority that motivated action among the participants within NOAA. NIDIS had a nationally recognized mandate to improve coordination around drought information and the resources to pull the partnership together. SCIPP, and its partner State Climatologists, had existing local connections that provided a conduit to stakeholders. All three were necessary components. Furthermore, collaboration with State Climatologists provided a meaningful local context for participants in the various activities and NDMC brought a wealth of prior experiencing in planning for, responding to, and managing drought. As U.S. Drought Monitor Authors, they were able to speak authoritatively on the drought status and outlook sections of the webinars.

NIDIS had an active interest, already funding several pilot projects in other parts of the country, and RISA had an interest through SCIPP and CLIMAS, which had been investigating drought issues for several years. However, without the coordinating role of the NOAA RCSD and the NWS CSPM, it is unlikely that such a concerted, multifaceted effort would have developed. The RCSD and CSPM initiated conversations with NIDIS, SCIPP and CLIMAS to develop the concept of the forums and webinars. Prior to the engagement, SCIPP had no experience conducting webinars and it is unlikely that such an effort would have developed independently.

Experience with the forums proved crucial to follow-on activities, both in subsequent forums across the region and in the format of the webinar series. Participants in the webinars found the status and outlook most useful, similar to anecdotal comments received during the forums. The NDMC-led status and outlooks were patterned directly upon the presentations in the morning sessions at the forums. The shortened briefings were subsequently based upon the slides used by NDMC in the status and outlooks portion of the full briefings. Each of these activities was evolutionary, based on the needs and involvement of stakeholders as the situation evolved.

SCIPP is premised on the ability to help communities respond to a range of extreme events. Through building these relationships in managing events as they happen, they build credibility and open the door to discussions on longer-term planning, including incorporating projections of climate change. So it was very much in the RISA team's interest to participate in the effort.

Participating in the forums and leading the webinars allowed SCIPP to expand its stakeholder connections. Evaluation of the webinars allowed an opportunity to examine how information was used. Thus, SCIPP was able to conduct research on the use of information among a group to which it may not have otherwise had as much access.

The one downside of participating in managing the real-time event is that the effort derailed some of the longer-term planning initiatives that had been launched in early 2011. In collaboration with CLIMAS, these response efforts focused on the most severely impacted areas of Texas, Oklahoma, and New Mexico. Consequently, less attention could be focused on the other four states in SCIPP's region, three of which do not have drought plans. Thus, it was difficult to maintain the connections in those states and little further progress has occurred with regard to drought planning in those states. As drought conditions have evolved more slowly since 2013, SCIPP and CLIMAS are again turning their attention to drought and climate adaptation planning and studying economic, ecological, and social impacts of the drought.

9.6   Participation by stakeholders

Because stakeholders' operational needs varied, multiple means of participating were used. While face-to-face meetings provided the best opportunity for interactions, they were time-consuming and expensive to develop. It required an investment of time for stakeholders to participate and was limited in spatial coverage. For the Southern Plains, the drought covered an area of approximately 400,000 square miles (the area of Texas, Oklahoma, and half of New Mexico). Assuming a "day-trip" travel radius of about 120 miles, each forum could serve an area of only 45,000 square miles. Thus, it would take nine evenly distributed forums to cover the full area affected by the drought.

The webinars offered a compromise, whereby stakeholders could participate in the presentations, ask questions, and even prioritize the content of future webinars. Their time commitment was much less—1 hour rather than 1 day—and no travel was required. Although webinars did not provide the opportunity for one-on-one conversations, they at least provided an opportunity for dialogue in a group setting, between stakeholders, researchers, and practitioners.

A survey was conducted in December 2011 on 250 people who had registered for the webinar series with SCIPP, with a response rate of 20% [13]. The drought conditions summary and outlook was rated highest (79% rated as an essential element). A majority of respondents (71%) indicated that they used one or more resources made available following the webinar, primarily the slides and the 2-page summary. Six respondents (12%) reportedly followed up with one or more of the contacts provided.

Successfully delivering information is only the first step in a hierarchy of standards of utilization [14]. As expected, the vast majority (79%) indicated that the webinars were useful for background information, but many respondents indicated other uses in addition. Particularly encouraging were the high response rates on "changed my perspective on an issue" (31%) and "raising my colleague's awareness of an issue" (46%). Eighty percent of respondents indicated that they had forwarded information to another person or other organizations. These indicate that the presentations were effective at these higher standards, beyond just background information or resources for what is already known.

Specific uses of the information were provided by 18 of the respondents. These were categorized as either operational decisions, use of materials in presentations, or raising awareness. Examples of how the webinar information was used, as cited by the respondents, included: assessing evaporative loss in management practices; influencing agency involvement in planning and operations; using information in public meetings; recruiting experts as presenters for other meetings; and using resources provided through the webinars.

Impact reporting was another way in which stakeholders could directly participate in the drought monitoring process. Information on how to submit impact reports was included during each webinar and briefing announcements and in mailings associated with the webinar. In addition, SCIPP used direct e-mail to solicit feedback from certain stakeholder groups, such as USDA field offices, cooperative extension, and water management districts, to diagnose conditions in areas where the extent of drought impacts was not fully captured by the Drought Monitor.

The addition of the briefings and postings on YouTube made it easy for people who only wanted a quick overview of drought conditions and outlooks without having to invest more than 5 minutes in the process. It also was a way to maintain the interest levels among the target demographic, in between more extended webinars or drought forums.

The use of technology allowed the climate services providers to reach a wider audience. Webinars made it possible to have a simultaneous conversation across a three-state region. The use of social media such as YouTube for posting video updates, in particular the short briefings, made explanations easily accessible to stakeholders on their own schedules. This increased the visibility for all service providers, with multiple points of entry for stakeholders to access the network. The long-term challenge is that new connections with stakeholders increases the demands for services and continuing these beyond the drought period will require sustained effort by all providers.

The providers also learned from the stakeholders how the information was used for decision-making. Information was not always used directly to make a decision but was used to bring awareness of the issue or as evidence why a decision was made. Indeed, the information seemed to have a high value in communicating the severity and exceptional nature of the event. Whether it was being used to influence a colleague's perspective of the drought, to initiate planning, or to communicate with the public, it was a way to establish a common understanding and possibly create flexibility for management decisions that could have negative consequences to the public and the local economy.

The process of the forums and webinars and the network that was created was also useful to the climate service providers and the decision-makers. The nascent network was used to recruit experts for other meetings and many of the webinar presenters were identified through the drought outlook forum process. Those connections are not continuously being used but likely could be re-established with less difficulty during the next drought.

9.7 Lessons learned

At the outset of the effort, the climate services providers assumed that their primary role would be to better link products and services with stakeholders. However, it quickly became apparent that discussions and interpretation of those products were more important in building confidence for stakeholders to accept the information and seek other information from the providers.

A key issue that arose in several of the venues was the need to facilitate use of the information, not just to provide that information. Stakeholders expressed interest in how to interpret graphics, the reasoning behind outlooks and assessments, and examples of how the information, products, or services are being used by others. This effort was an exchange between researchers and decision makers because researchers and operational meteorologists were interested in learning when products and information were needed to make critical decisions. For example, many water allocation decisions are made during late winter or early spring, before the outcome of the primary rainy season is known. Decisions on planting and insurance could benefit from more use of seasonal forecasts, but the critical time window for those decisions corresponds with a time when seasonal forecasts have their lowest skill. These dialogues helped each group appreciate the constraints under which the other group operates, even if solutions could not be developed (for example, forecast skill could not be improved to benefit planting decisions).

Another facet of the discussions is that stakeholders were more interested in explanations of science and processes rather than management advice. Because the primary stakeholders were part of an existing community of practice [15,16], for example professional cattle associations, they already had access to trusted sources of information on management practices. What was newer to them was the scientific basis of the causes, physical inter-relationships, and processes related to drought. Consequently, topics such as how Pacific Ocean sea surface temperatures could influence the Southern Plains climate, the interaction between surface water and ground water, and how the U.S. Drought Monitor maps are created each week were among the most popular topics. The webinars and forums provided an opportunity for scientists and operational meteorologists to share their expertise and discuss developmental products (such as an evaporative stress index) with a diverse audience of practitioners.

Through these interactions, stakeholders were looking to gain confidence in the forecast, learn about limitations of predictability, and examine ways of applying those forecasts to their operations. The greatest challenge is that the scale of decision-making—seasonal to interannual—is the period when meteorologists and climatologists have the least skill in making predictions. Consequently, the general message of these interactions was that, based on a set of causative factors, these were our best estimates for the upcoming season to year, but also understanding that there is a rather broad range of scenarios that are nearly equally plausible.

These discussions helped to build trust that resulted in increases in receptivity of the forecasts and outlooks. This was borne out in the experiences with the 2012 La Niña and 2013 El Niño events. During 2011, the provider team had promoted the message of a second-year La Niña; however, the winter and early spring unexpectedly turned wet across eastern parts of the region, contradicting the fall outlook. Despite the poor forecast, there was no discernible decline in participation in the forums or webinars. The following summer, the message was of a developing El Niño signal, which favored a wet pattern across the region. El Niño fizzled and the following seasons remained dry, except in central Oklahoma. Yet, participation in the forums, webinars and added briefings remained high. Among factors determining the use of scientific and technical information is a desire for certainty [17]. Even though the outlooks had a high degree of uncertainty, the engagement process seemed to counteract the desire for certainty heuristics and make participants more willing to consider the highly uncertain nature of the outlooks in their planning decisions.

In placing products and information in the context of their decision-making processes, new or refined research needs were also identified. Some of the topics that require further study are the ecological impacts of drought, wildfire climatology, changes in fire regimes with a warmer and drier climate, quantifying the evaporative losses from reservoirs and vegetation, and creating an integrated reservoir database that will place reservoir behavior within its historical context. Initiatives were developed in all of these areas with some new products and services already under development.

Another area that needs further research is the importance of differing impacts on different time scales. In weekly Drought Monitor discussions and comparing impacts to indices, it was often difficult to determine what time scales should take precedence in assigning the appropriate drought category. Short-term rainfall can mask long-term deficits, such that the surface water is quickly absorbed deep into the soil or evaporated back to the atmosphere. Some of these processes are not adequately reflected in drought indicators, providing a potential area for further study.

Scale also applies to the services provided (see Chapter 11). One of the challenges is determining an appropriate area of coverage without it being so large that information is diluted and not highly relevant to stakeholder decisions or being so small that it applies to only a small segment of the affected population. In addition, the providers had to contend with jurisdictional boundaries, primarily state lines. Many existing professional and social networks are established on the basis of state associations, so generating a dialogue between state entities was a priority for the climate services provider team. Webinars endeavored to include presenters from multiple states so that comparisons of management practices were possible.

Lastly, this was more of a process to build a community than a product. Providers and stakeholders both learned from experience along the way and through iteration improved the information provided and illuminated the context in which decisions were made. While in many instances the response to the drought has remained reactive, these discussions have yielded an abundance of information that will form subsequent development of best practices guidelines, improve drought planning, and connect state and local monitoring more closely. As drought conditions abate, it will allow partners such as SCIPP and CLIMAS to continue interaction with these stakeholders to examine longer-term adaptation and planning for drought and other climate-related impacts.

9.8 Summary

The drought that began in 2011 provided extreme and long-lasting challenges for the Southern Plains. Through engagement with stakeholders the region's climate services providers were able to deliver products, discuss causes, and improve monitoring. Through the process of meetings, forums, webinars, and monitoring, several lessons were learned. Delivering products is not sufficient; interpretation, discussion of reasoning behind forecasts, and explanation of physical processes enabled stakeholders to have more confidence in use of products and services. Decision-makers already know their options and management strategies; what the climate services partners were able to bring to the table was an integration of seasonal climate information into these ongoing decision processes.

Discussions and interactions helped to build trust and increase receptivity of the information. Through these bi-lateral discussions, providers were able to identify research needs and scope new products to meet decision needs. Stakeholders not only used the information for their own decisions, but also used it to influence others. Scale of services emerged as an important factor. Decision-makers operated on different scales, from the ranch to the region. Some products were better suited for different scales than others. The drought engagement efforts provided an opportunity to work on all of these scales and identify gaps in products and services at some scales that may not have been as apparent at other scales.

Through a coordination of efforts, initiatives were launched to reach a wide audience affected by the drought and to maintain their interest throughout the years. The combination of forums, webinars, and planning and monitoring provided multiple points-of-entry into the collective drought monitoring and management system. These multiple pathways to and from science helped to build an ongoing community around managing drought. Each of these efforts has met with some success, but working together they have reinforced each other and sustained a dialogue on drought that extends not just to managing the current challenges, but also in taking a longer-term perspective on a disaster as insidious and difficult to describe as drought.

References

1 Bauman, A., Goemans, C., Pritchett, J. *et al.* (2013) Estimating the economic and social impacts from drought in southern Colorado. *Journal of Contemporary Water Research & Education*, **151**, 61–69.

2 109th Congress (2006) *Public Law 109-430, National Integrated Drought Information System Act of 2006*. URL http://www.gpo.gov/fdsys/pkg/PLAW-109publ430/html/PLAW-109publ430 .htm [accessed on 22 July 2013].

3 National Climatic Data Center (2012) *Billion Dollar U.S. Weather/Climate Disasters 1980–2011*. URL http://www.ncdc.noaa.gov/billions/events.pdf [accessed on 9 September 2015].

4 Seager, R., Kushnir, Y., Herweijer, C. *et al.* (2005) Modeling of tropical forcing of persistent droughts and pluvials over western North America: 1856–2000. *Journal of Climate*, **18**, 4068–4091.

5 Nielsen-Gammon, J.W. (2011) *The 2011 Texas Drought: A Briefing Packet for the Texas Legislature*, October 31, 2011, pp. 43. URL http://climatexas.tamu.edu/files/2011_drought.pdf [accessed on 9 September 2015].

6 Peterson, T.C., Stott, P.A. & Herring, S. (2012) Explaining extreme events of 2011 from a climate perspective. *Bulletin of the American Meteorological Society*, **93** (**7**), 1041–1067.

7 Stone, D. (1996) *Capturing the Political Imagination: Think Tanks and the Policy Process*. Frank Cass & Co., London.

8 Wilhite, D.A., Hayes, M.J., Knutson, C. *et al.* (2000) Planning for drought: moving from crisis to risk management. *JAWRA Journal of the American Water Resources Association*, **36**, 697–710. doi:10.1111/j.1752-1688.2000.tb04299.x

9 Emery, M., Gutierrez-Montes, I., Fernandez-Baca, E. (eds) 2013. *Sustainable Rural Development: Sustainable livelihoods and the Community Capitals Framework*, pp. 120. Routledge.

10 Pulwarty R.S., Sivakumar M.V.K. (2014) Information systems in a changing climate: early warnings and drought risk management, *Weather and Climate Extremes* **3**, 14–21. 10.1016/j.wace.2014.03.005.

11 Lackstrom, K., Brennan, A., Ferguson, D. *et al.* (2013) *The Missing Piece: Drought Impacts Monitoring*. Workshop report produced by the Carolinas Integrated Sciences & Assessments program and the Climate Assessment for the Southwest.

12 Brabham, D.C. (2008) Crowdsourcing as a model for problem solving: an introduction and cases. *Convergence: The International Journal of Research into New Media Technologies*, **14**, 75–90.

13 Shafer, M. & Riley, R.. (2012) *Managing Drought in the Southern Plains: A Summary of Survey Responses to the Webinar Series*, pp. 10. Southern climate Impacts Planning Program, Norman, OK. URL http://www.southernclimate.org/publications.php [accessed on 9 September 2015].

14 Knott, J. & Wildavsky, A. (1980) If dissemination is the solution, what is the problem? *Knowledge: Creation, Diffusion Utilization*, **1**, 421–442.

15 Brown, J.S. & Dugid, P. (1991) Organizational learning and communities-of-practice: toward a unified view of working, learning and innovation. *Organization Science*, **2** (**1**), 40–57.

16 Hildreth, P. & Kimble, C. (2004) *Knowledge Networks: Innovation Through Communities of Practice*. Idea Group Publishing, Hershey, PA.

17 Slovic, P., Fischoff, B. & Lichtenstein, S. (2010) Rating the risks. *Environment: Science and Policy for Sustainable Development*, **21** (**3**), 14–39.

SECTION IV

Advancing science policy

Advancing Science Policy, the final theme in *Climate in Context*, addresses the ways by which the norms and culture of science need to adapt, in order to address emerging societal challenges. "Science policy" refers to the decision processes through which individuals and organizations support, manage, and evaluate research. This means that science policy plays out at many levels, from decisions that scientists make about publications, research methods, and grant proposals to high-level budget allocations of federal agencies, Congress, or private companies and foundations. Although it is often framed as a budget problem, science policy is also about the culture of science—the incentives and inertia built into a complex system of institutions.

The Regional Integrated Sciences and Assessments (RISA) program, by itself a product of innovative science policy, has created a space to learn about, and experiment with, the different approaches to connecting science with users. Doing away with the traditional 3-year grant evaluated almost entirely on technical merit, the RISA program elevates user needs, partnerships, societal outcomes, and problem-based approaches alongside narrower scientific concerns. This represents a change in the way that federal managers overseeing the program conduct the business of funding science. However, it also gives scientists an opportunity to make decisions differently, think differently about their goals and their practice, and thereby impact the culture of science.

RISA teams have demonstrated the importance of flexible governance for responding to factors that motivate interactions between scientists and decision-makers, which include, among others, natural disasters, institutional change, climate literacy, and breakthroughs in science. They maintain diverse structures for program leadership and management, which is critical for maintaining the healthy relationship between multiple institutions, as well as for leveraging scientific capabilities within regions. Furthermore, they devote resources to, and develop expertise in, the specific challenge of linking between scientists and decision-makers.

Climate in Context: Science and Society Partnering for Adaptation, First Edition.
Edited by Adam S. Parris, Gregg M. Garfin, Kirstin Dow, Ryan Meyer, and Sarah L. Close.
© 2016 John Wiley & Sons, Ltd. Published 2016 by John Wiley & Sons, Ltd.

Because of this innovative program-building, the various manifestations of RISA programs have also raised important questions about science careers, and the institutions that support them. As scientists have ventured beyond their intellectual and practical comfort zones, RISAs have confronted the challenges of evaluating this work within institutions (e.g., research universities, federal funding for basic science) that traditionally focus on simple indicators such as journal publications.

Chapters throughout this volume are related to science policy: they synthesize what the National Oceanic and Atmospheric Administration (NOAA) and its grantees have learned about making RISAs successful. They also show that these lessons are valuable far beyond the confines of NOAA, or the topic of climate. The norms of science are changing, and more and more scientists are pursuing use-inspired, problem-driven research. RISAs provide an opportunity to ask big questions about the challenges and opportunities that come with these changes.

In Advancing Science Policy, Ferguson *et al.* make the case for using a programmatic theory of action, in conjunction with well-articulated, pre-defined measures, to assess the value and effectiveness of use-inspired collaborative research. They caution, however, that it is often difficult to establish causal mechanisms between interventions, such as the introduction of forecasts, information, and decision support, and the outcomes in addressing real-world societal challenges, that is, the type generated by exposure to climate variability and change and the underlying vulnerabilities of social and ecological systems. Gordon and colleagues look at the science–policy interface from a geographic and institutional perspective, and the ways in which they intersect with the scope of RISA activities. They point out that the relationship building and convening functions of boundary organizations, such as RISAs, require greater investments of time, and a new set of incentives that rewards these activities, in addition to traditional research. These cases point to valuable lessons that RISAs have learned for the program as a whole, and for the broader science policy community striving to derive public value from investments in research.

CHAPTER 10

Evaluation to advance science policy: lessons from Pacific RISA and CLIMAS

Daniel B. Ferguson[1], Melissa L. Finucane[2,3], Victoria W. Keener[2] and Gigi Owen[1]

[1] *Institute of the Environment, University of Arizona, 1064 E. Lowell Street, Tucson, AZ 85721, USA*
[2] *East West Center, 1601 East-West Rd, Honolulu, HI 96848, USA*
[3] *RAND Corporation, 4570 Fifth Ave #600, Pittsburgh, PA 15213, USA*

10.1 Introduction

The regional partnerships developed through the RISA program have been highlighted as effective institutions for providing climate information and decision-support services to diverse stakeholders [1]. The RISA teams have been described as innovative in "organizing the dialogue between scientists and practitioners" [2] to generate science "that is usable in specific decision contexts" [3]. Unfortunately, while there is ample anecdotal evidence to support these assertions, there is limited empirical work to demonstrate the impacts of regional RISA programs or the overall impact of the national RISA program. Thus, few robust conclusions can be made about how and why RISA research and outreach does or does not support societal decision-making and climate adaptation planning.

The purpose of this chapter is to explain why it is important for RISAs to undertake evaluations and to provide examples of metrics and methods for evaluating programs implemented in a complex, real-world environment. We discuss evaluation activities by two RISA teams—Pacific RISA and the Climate Assessment for the Southwest (CLIMAS). We provide a simple framework for connecting the RISA program evaluation to science policy at multiple scales, and illustrate with examples the evaluation undertaken by Pacific RISA and CLIMAS. We argue that to inform science policy

Climate in Context: Science and Society Partnering for Adaptation, First Edition.
Edited by Adam S. Parris, Gregg M. Garfin, Kirstin Dow, Ryan Meyer, and Sarah L. Close.
© 2016 John Wiley & Sons, Ltd. Published 2016 by John Wiley & Sons, Ltd.

across scales, we need to design evaluations so that results are meaningful and legitimate, and, at the same time, also allow for the highly iterative and adaptive nature of the environments in which RISA work is done and used. We conclude the chapter by distilling the lessons learned from the evaluation initiatives undertaken thus far.

10.2 The importance and challenge of evaluating RISA regional programs

Evaluation of well-designed RISA regional programs can provide valuable information regarding both *policy for science* (i.e., considerations of priorities for research funding) and *science for policy* (i.e., the utilization of research in decision processes) [4]. Regarding policy for science, evaluations provide information to science funding agencies, researchers, stakeholders, and the public about promising avenues and potential pitfalls of particular areas of research. Establishing evidence of the mechanisms through which RISAs are effective can bolster arguments for funding and prioritizing decision-relevant science. Similarly, evaluations provide RISA teams and their leaders with information on how to support program-level decision-making and thus improve program performance. Interim evaluations may be used to support course-corrections that improve RISA focus, methods, and outcomes. Regarding policy for science, evaluations help to advance the emerging field of participatory climate research by demonstrating how research is being used and incorporated into decision-making. Robust evaluations that provide reliable and valid findings yield generalizable knowledge about how climate science can be used for effective decision-making in different contexts for different stakeholders.

The central challenge for evaluation research is to understand the value of a program by causally linking program activities with outcomes and impacts that can be attributed to that program. In most applied research, three criteria must be met to establish causality: (1) temporal precedence (did the cause happen before the effect?); (2) covariation (is the effect observed when the cause is implemented?); and (3) no plausible alternative explanations (is a third factor causing the outcome?). Ruling out alternative explanations is the most difficult criteria to address in a RISA program evaluation because the complexity and real-world nature of RISA activities—and the iterative and adaptive processes used—are incompatible with tightly controlled experimental designs. RISA activities are deliberately dynamic and responsive to changing needs, opportunities, and challenges. Although each RISA has a mission and its own strategic goals, the very nature of doing participatory and transdisciplinary[1] research means "that the intended impacts are likely to change as different people become engaged, opportunities and threats emerge, and the program unfolds" [13, p. 77]. Many factors may be changing simultaneously, making it hard to pinpoint exactly what is causing the observed effects. In this kind of a complex environment, it is nearly impossible to predict emergent outcomes (either intermediate or long term) with any precision.

For instance, researchers may observe that stakeholders pay more attention to climate projections after they are engaged in a series of dialogues, training workshops, and tool development. However, isolating the effect(s) of any one (or set) of these activities may be very difficult given the multipronged approach of a RISA team. Researchers would also need to collect and compare measures of the extent to which stakeholders pay attention to and/or understand information at multiple points in time (e.g., before and after each activity). However, stakeholder fatigue may not make this possible and small sample sizes may render statistical analyses of such data inappropriate. Additionally, effects from changes in the external environment (e.g., a directive from leadership in a stakeholders' organization to consider climate change impacts in adaptation planning) may be impossible to capture precisely given that we cannot control the socio-political context in which a program is being implemented.

Despite the above challenges, we need to be able to learn about the value of RISA programs, even without establishing causality. In the remainder of the chapter, we discuss how evaluation can be done in complex, real-world environments and how the lessons learned are still useful even when direct causality has not been established. We organize our discussion below to address first the need to inform decisions about science funding priorities (policy for science). Then we address the need to facilitate the use of research results in decision processes (science for policy).

10.3 Policy for science

Science is expected to address increasing numbers of societal problems within shrinking research budgets. Decisions about what science should be prioritized in this context are difficult. A joint memo from the Office

---

[1]Following Jahn *et al.* [5], we define transdisciplinarity as "a critical and self-reflexive research approach that relates societal with scientific problems; it produces new knowledge by integrating different scientific and extra-scientific insights; its aim is to contribute to both societal and scientific progress".

of Management and Budget and the White House Office of Science and Technology Policy to the heads of executive departments and agencies in 2009 states that: "Agencies should develop outcome-oriented goals for their science and technology activities, establish procedures and timelines for evaluating the performance of these activities, and target investments toward high-performing programs. Agencies should develop 'science of science policy' tools that can improve management of their research and development portfolios and better assess the impact of their science and technology investments" [6]. In other words, all U.S. programs that fund research need to clearly articulate how investments can return value.

The challenge—and one that the RISA teams are well-positioned to address—is to develop broad indicators and metrics of societal value and to monitor those that allow funders and policy makers to assess the investment in publicly funded science beyond traditional economic measures. This knowledge is critical for funders and politicians who must make crucial policy decisions about where to invest research funds to best "foster societal well-being" [7]. In addition, this knowledge supports leadership within RISA teams to make decisions about priorities for new work and allocation of scarce resources.

Obviously, any individual RISA evaluation effort will make modest progress toward these high-level directives, but these efforts—especially if every RISA incorporates evaluation into their program activities—can offer critical insights into National Oceanic and Atmospheric Administration (NOAA) and other funders who seek to support the kind of use-inspired research conducted by RISA teams. At the most basic level, RISA evaluations can develop metrics for assessing this work that reflect the inherent nuances. While standard economic metrics are often sought to justify research funding (e.g., return on investment), these metrics often fail to reflect the actual impact of RISA research and outreach. Similarly, traditional measures of the value of scientific knowledge (e.g., number of peer-reviewed publications) may reveal very little about the utility of that knowledge for addressing a complex problem like vulnerability to climate variability or change. Despite the current pressure from high-level policy makers, "the capacity to evaluate, either prospectively or even retrospectively, the potential for particular research priorities and institutional arrangements to achieve stipulated *noneconomic goals and values* remains primitive at best" (emphasis added; [7]). This is a critical component of the nascent field of the science of science policy that the RISA evaluations can help advance.

10.4   The role of theory-based evaluation

Several reviews describe a range of approaches that are available for program evaluation and how the appropriate approach depends on the program to be evaluated [8–12]. The intellectual basis for current Pacific RISA and CLIMAS evaluation efforts is theory-based evaluation (see [13, Chapter 2]). Theory-based evaluations have become the dominant approach for many policy-to-action programs because of their inherent flexibility.

The core of theory-based program evaluation provides a well-defined description of how program activities lead to particular outcomes. In simplest terms, the program theory outlines the logic of how a program operates. The critical component of a program theory for evaluation is what is called a theory of change [13] or an impact model [8] that has been independently tested and empirically verified. The program theory is useful for identifying the hypotheses about the linkages between a program and its outcomes and impacts, which basically is what the evaluation aims to test. Researchers can then choose specific indicators and methods that are appropriate for testing these hypotheses.

Opinions vary on the appropriate role of a program theory in evaluation (e.g., see criticisms by [14]). One limitation of the program theory, as mentioned above, is that causality may be hard to establish. A second limitation is that if the theory is incorrect or poorly articulated, then the evaluation will yield little useful information. And finally, although program theory is widely discussed in methodology literature, many evaluations do not yet have an accompanying theory of change. Consequently, instead of using fully developed program theories, researchers may use theories of action (or theories of implementation), typically presented in an Action-Logic Model (ALM).

Logic models help to conceptualize, identify, and implement a comprehensive range of relevant evaluation metrics at many steps along the reasoning chain, which allows us to examine the multiple steps in the underlying reasoning of a program. Examining the process as well as the outputs and outcomes allows a well-designed evaluation to demonstrate faulty assumptions or weaknesses in the reasoning that may be responsible for failure to achieve the expected outcomes. This feature of logic models creates the potential for the monitoring and adaptive management of the different aspects of the program to both improve performance and uncover emergent phenomena that may have been absent from the initial program theory.

Given the complex contexts in which RISA programs are implemented, logic models provide an appropriate way to organize evaluation activities when establishing causality is problematic.

The evaluation examples provided below reflect different stages of adoption of the principles of theory-based evaluation. The Pacific RISA team has developed an ALM as the centerpiece for their evaluation effort (details below). The CLIMAS team, on the other hand, undertook several years of project-based evaluations that were not organized around a central logic model, but has recently begun to use a theory-based approach to ongoing program-level evaluation.

10.5 Conceptualizing how change occurs: developing an Action-Logic Model

In its first year of funding, the Pacific RISA team developed an ALM to guide and assess the effectiveness of all research projects. Figure 10.1 graphically illustrates the ALM, highlighting the connection of six interdependent components in the system. Like a road map, the ALM shows the steps taken to achieve the long-term goal of building climate-resilient island communities. The starting point is the *context* or setting (the sociopolitical, environmental, and economic conditions) in which the program objectives address the mission of the Pacific RISA. The context also identifies the factors considered when priorities are set to direct the program. Understanding the rich context in which the program is implemented is necessary to know what resources and activities are most appropriate to achieve the long-term goal.

Next, *inputs* are identified, which include the resources needed (human, financial, or other) for program activities to build capacities that address the long-term goal. The *outputs* include the activities conducted as well as the participation by the target audiences. The outputs essentially describe what the program does with the resources to direct the course of change. We expect that specific activities (e.g., scientific research, workshops, partnering) will provide diverse ways to understand what information and support is needed (and technically possible) for various audiences. By reaching certain individuals and groups, certain *outcomes/impacts* are expected to be achieved, occurring over time from short- to medium- to long-term. In the first 1–2 years we expect to see increased awareness of climate–society interactions; increased knowledge of risks, vulnerabilities, and needs; and improved skills in using climate science. Then, with additional investments and activities in years 3–4, we expect to see changes in behaviors, practices, procedures, and policies. Ultimately, in about 5 years or more, we expect

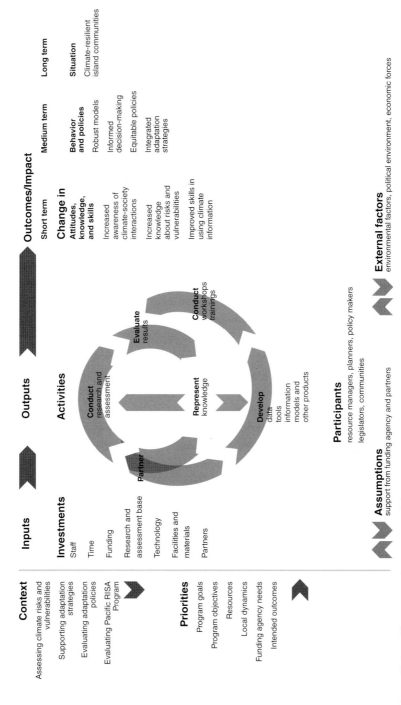

Figure 10.1 Pacific RISA Action-Logic Model.

to see a change in the situation such that island communities are more climate-resilient.

The ALM also identifies *assumptions*, which are underlying beliefs about the way members of the team think the program will work. For instance, we assume continued support from funders and partners because without this the nature and extent of activities that can be addressed would be much more limited, with a concomitant change in outcomes and impacts. *External factors* include conditions beyond the direct control of the Pacific RISA that nonetheless influence the program's success. For instance, changing environmental conditions, such as the development of extreme drought conditions will likely lead program participants to be better served by interannual precipitation and drought forecasts more than end-of-century sea-level-rise projections.

Finally, the ALM also shows arrows between and within the components, indicating that review and adjustment are ongoing, iterative processes—both in initiating the program and developing the model of change. This process reflects the essential "translational" role of RISA programs in converting climate science into a change in attitudes and behaviors.

10.6   Conceptualizing how change occurs: understanding outside perspectives

In contrast to the Pacific RISA approach, CLIMAS did not begin their evaluation activities by developing an ALM. Rather, in 2007–2008, CLIMAS researchers and a small group of external evaluators undertook a project that aimed to better understand how stakeholders perceived the program and to document the ways they interacted with it. That information has influenced programmatic decision-making (e.g., amount of effort to expend on different categories of activity) and has directly shaped the way that CLIMAS leadership and team members conceptualized the program's role in the regional climate services network. One of the most salient results from that project was a conceptual model of how stakeholders in the Southwest conceived of their relationships with CLIMAS. Four primary roles that CLIMAS was perceived to play were identified: (1) as an information broker communicating climate information (e.g., climate summary publications, conference presentations); (2) as an informal consultant (e.g., providing expert advice on project development, data access, and/or analysis); (3) as a short-term partner (e.g., cosponsoring an event); and (4) as a collaborator where long-term relationships were established to work through a series of issues relevant to both CLIMAS and the partner (Figure 10.2).

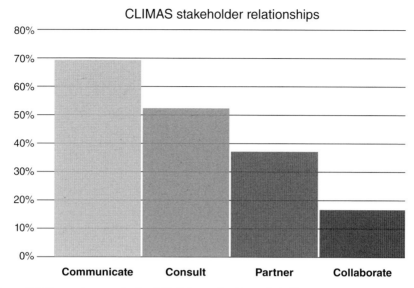

Figure 10.2 Four primary roles for CLIMAS perceived by stakeholders.

When given a list of four categories for ways they may interact with CLIMAS and asked to choose all that applied, 70% of respondents reported that the program acted as an information broker, 53% said they consulted with CLIMAS for expert advice, 37% said they partnered with a CLIMAS researcher for a short-term project, and 17% said they considered themselves collaborators with CLIMAS researchers. Aside from helping to improve CLIMAS' conceptualization of its role in the Southwest, this project revealed new research and evaluation questions that need to be explored. For example, to what extent is the amount of time someone interacts with a CLIMAS researcher the key determinant in whether or not they ultimately partner or collaborate with CLIMAS? Metrics that capture that temporal dimension—now a component of ongoing CLIMAS evaluation activities—will help support or refute the theoretical claim that engaged climate stakeholder work takes sustained, long-term interactions, a key theme in ongoing science policy conversations about best practices for participatory, problem-oriented climate research.

10.7 Metrics and learning from self-evaluation plans

Despite their different approaches to conceptualize how change occurs, both Pacific RISA and CLIMAS have developed self-evaluation plans to track their roles in the process of change. These plans include metrics that incorporate, but go beyond, our funder's annual reporting requirements. This information is useful at the scale of our regional programs for assessing and prioritizing current and future research needs, but is also useful for NOAA at the national level because it provides on-the-ground indications of progress that can help them better understand how their investments are (or are not) meeting program goals. Robust self-evaluation can therefore provide a wealth of useful information to guide science policy decision-making at the level of a funder or even across a range of federal agencies.

To illustrate, at the beginning of the current funding cycle, the Pacific RISA leadership (with assistance from the program's external evaluator) developed a comprehensive self-evaluation plan, which included: (1) a set of potential qualitative and quantitative metrics to evaluate program objectives in the near-, mid-, and long-term; (2) three areas of research and assessment to quantitatively evaluate; (3) a project progress tracking sheet for team members to complete on an annual basis; (4) a commitment to evaluate each workshop that Pacific RISA organizes; and (5) an internal website to track internal and external factors that may impact the program objectives. Guided by the report of the National Research Council's Committee on Metrics for Global Change Research [15], the goal in choosing the final set of Pacific RISA evaluation metrics was that they address the quality of program leadership and strategic planning, foster future progress, and be useful for independent, transparent peer review (see Table 10.1 for examples).

Over the past several years, CLIMAS has undertaken a similar process of internal tracking. The current CLIMAS tracking system is now tied directly to four key areas central to the ongoing program evaluation: outreach, graduate student education and training, decision support, and creation of new knowledge. Using a series of metrics tied to these broad categories, CLIMAS is gathering data and information from each member of the CLIMAS team several times a year. This information is annually synthesized for reporting to NOAA and also to provide input to team leadership for setting research priorities.

By tracking project progress through annual reporting, monitoring internal and external impacts to the program, and measuring project progress as part of a normal review and planning process, both Pacific RISA and CLIMAS continually evaluate our evolving programs. When combined with our externally focused evaluation efforts, both of our programs are providing—annually as well as at the end of each 5-year funding cycle—specific information to demonstrate how our programs have met our objectives. In this way, policy for science is enhanced through state-of-the-art knowledge about what set of factors most effectively increase the production and use of climate science to address complex societal challenges.

Table 10.1 Examples of short-term, long-term, qualitative, and quantitative metrics used in the Pacific RISA program evaluation process.

Component of action-logic model	Variable or indicator	Metric
Context/rationale		
Assessing climate risks	– Scientific understanding – Practical experience	– Model outputs – Qualitative observation – Quantitative change in knowledge via survey response
Inputs		
Financial support as planned from NOAA	– Level and continuity of support from funding agency	– Funding amount requested and received – Expected date and actual date of funding
Outputs		
Workshop research activities	– Interest among stakeholders – Learning and change in knowledge	– Attendance and feedback from post-workshop evaluations – Expressed feedback on learning impacts
Research and assessments	– Research conducted – Key findings and novel insights – Presentation of findings	– Peer-reviewed publications and other reports – Downloads of publications or website visits – Media coverage generated
Partnerships and collaborations	– Degree, type, and quality of partnership	– Lists of partners and stakeholders – Description of roles and involvement
External factors		
Progress of state or county adaptation planning	– Type of and/or change in adaptation planning activity	– Existing or planned adaptation plans – Executive or Gubernatorial orders – Regulatory changes
Environmental factors	– Climate-related extreme events and disasters – Nonclimatic environmental problems	– Disaster impacts – Other event impacts (financial, change in public support)
Outcomes		
Short-term (1–2 years)	– Changes in stakeholder knowledge or awareness – Level of trust between scientists, stakeholders, and among partners	– Self-reported perceptions of climate change importance – Change in reported attitudes – Quality of interactions and self-reported trust

10.8 Science for policy

The RISA program is designed to "help expand and build the nation's capacity to prepare for and adapt to climate variability and change" [16]. Following that mission, each of the RISA teams carries out research and outreach relevant to regional decision-making. The nature of RISA projects, therefore, provides an opportunity to assess the extent to which research supports public policy as well as understands how these decision contexts influence the research we carry out. By developing processes and projects that are designed to gather data and information about how research is being incorporated into decision-making, RISA evaluations shed light on the complex ways that science and public policy interact. This knowledge is critical for policy makers who must make difficult decisions about how to address societal needs effectively.

Both Pacific RISA and CLIMAS evaluation projects include efforts to analyze interactions between the research we conduct and its use in decision-making. Ideally, by better understanding how our research is being used, we become more adept at generating "knowledge that [is] useful and valuable for the broad set of institutions that … find themselves in the position of confronting various aspects of the climate problem" [7]. Thus, RISA evaluations become a mechanism for more integrated and refined science for policy—at least at the scale of our individual programs—that is responsive to the decision-making needs in our regions. In the examples in the following text, we highlight how evaluation efforts demonstrate the ways our research is directly used in policy making.

10.9 Informing policy: analysis of the Hāwai'i Water Code

Evaluation of a completed Pacific RISA project done in conjunction with partners at the University of Hāwai'i Center for Island Climate Adaptation and Policy (ICAP) demonstrates one way in which research has led to ongoing policy-relevant outcomes. In the ALM's context of "evaluating adaptation plans and policies," a Pacific RISA legal scholar analyzed the law and policy framework of Hāwai'i to identify ways of enhancing climate adaptation for the freshwater resources of Hāwai'i without introducing new legislation.

Methods included a review of recent case studies and peer-reviewed literature on adaptive governance. The work resulted in a white paper, [17] which describes climate change risks to the water resources of Hāwai'i and principles of adaptation; analyzes the existing water law and policy framework of Hāwai'i based on four criteria that define the "adaptive capacity" of freshwater laws and policies (e.g., forward looking, flexible, integrative, iterative); and provides 12 tools (recommendations) to better adapt to climate change. The State of Hāwai'i Commission on Water Resource Management (CWRM) has started implementing several of the white paper's tools: enforcing regular updates to the Hāwai'i Water Plan and including climate change impacts; tying water use permitting fees more closely to costs; and adopting climate-conscious and forward looking groundwater sustainable yields (SY)—the theoretical rate at which groundwater may be pumped from a well without impairing the ability of that aquifer to provide enough water of the desired quality. Specifically, CWRM has hired a contractor to update the state Water Resources Protection Plan[2]; is working with Pacific RISA to create future climate and water management scenarios on Maui; and has drafted administrative rules that will allow fee flexibility (July 2013, personal communication, CWRM).

In May 2012, at the same meeting that CWRM Commissioners were briefed on the white paper, Monsanto applied for a water use permit for allocation of additional potable groundwater[3] as a "Future Emergency Back-up." Although a review of existing allocations indicated that this was under the SY, CWRM denied Monsanto's request, evidently incorporating concepts from the white paper into the arguments. Commission staff recommended that providing two allocations for the same use was an "unsupportable approach to scarce natural water resources"[4] amounting to "banking" water, and that CWRM denies the water use permit, but approves the well and pump permit subject to Monsanto submitting an emergency plan that would outline the conditions under which they would need it. In December 2012, the Commission adopted the staff's recommendation.

CWRM's approach in assessing Monsanto's request illustrates three adaptive principles from the white paper. First, rather than engaging in a rote review of SY, the analysis considered forward-looking impacts. Second, the decision incorporated flexibility for CWRM to address climate and future issues as they arise, in contrast to Monsanto's request, which would have resulted in a fixed allocation into the indefinite future. Finally, the

---

[2] *See* Minutes, December 19, 2012 Meeting of the Commission on Water Resource Management, *available at* http://hawaii.gov/dlnr/cwrm/minute/2012/mn20121219.pdf (emphasis added).
[3] Monsanto is a large player in the seed crop industry of Hāwai'i. The State Water Code includes a policy preference for agricultural water use. *See* Haw. Rev. Stat. 174(c)(2).
[4] *See* Commission on Water Resource Management Staff Submittal, Monsanto Company Applications for: (1) Ground Water Use Permit; (2) Well Construction Permit; and (3) Pump Installation Permit, 6, 9 (15 August 2012), *available at* http://hawaii.gov/dlnr/cwrm/submittal/2012/sb201208D1.pdf.

decision illustrates an integrated approach. Although Monsanto's request was for allocations from two separate sources, the decision considered those allocations in aggregate. These decisions are examples of short-term impacts (increased awareness of climate risks) and medium-term impacts (informed decision-making, integrated strategies) of Pacific RISA on public policy and decision processes.

10.10   Expanding capacity: needs, applications, and effective means for delivering climate information

Another way in which RISA evaluation activities help to expand the nation's ability to address the impacts of variable and changing climate is to systematically examine the mechanisms by which RISA teams effectively deliver climate information. The following text describes qualitative and quantitative data that offer an insight into how RISA programs have developed a reputation as a reliable source of regional climate research and information and contextualized information based on characteristics of users and climatic events specific to their region.

10.10.1   Building CLIMAS as a reliable source of information

In 2008, CLIMAS researchers carried out an evaluation of the monthly *Southwest Climate Outlook* (*SWCO*) regional climate summary, which aimed to improve the publication by ensuring that the information was relevant, useful, and presented in a way that met readers' needs. Through an online survey of readers and follow-up interviews with a subset of respondents, the evaluation revealed that *SWCO* often provided the respondents' first interaction with CLIMAS and served as a way to build the program's reputation as a reliable source of climate information. Even though this evaluation was geared toward simple improvement of the product, it helped the team conceptualize *SWCO* as: a critical gateway to the larger CLIMAS program; a platform to deliver current and forecast conditions and short essays on state-of-the-art developments in climate research; and a "boundary object" [18,19] that is meaningful for both CLIMAS team members and regional stakeholders.

10.10.2   Contextualizing information

In 2011, CLIMAS evaluated the experimental periodic Web publication called *La Niña Drought Tracker*, which interpreted drought and climate information for Arizona and New Mexico during the back-to-back La Niña years of 2010–2011. The *Tracker* evaluation tested the hypothesis that synthesizing regionally specific climate information would be useful during the extreme conditions that often accompany La Niña in the Southwest. Of 117 respondents to an online survey, more than 90% reported that their understanding of climate and drought improved by reading the *Tracker*. Approximately 67% of respondents said they used the *Tracker* to make a drought-related decision; of these, 21% reported that the *Tracker* played a significant role and 36% said it played a moderate role (for full results, see [20]). The *Tracker* evaluation showed the utility of brief, synthesized information focused on a single climate hazard (drought) that came from a trusted source.

10.10.3   Refining information delivery methods

Between 2009 and 2011, CLIMAS collaborated with RISA colleagues from the California Nevada Applications Program and the Alaska Center for Climate Assessment and Policy to investigate how seasonal climate and monthly fire potential outlooks impacted decision-making in wildfire management. In the Southwest, we conducted semistructured interviews with 37 wildfire management professionals and an abbreviated online survey with 40 members of the wildfire community. Some questions provided data for a social network analysis, which demonstrated that person-to-person communication with meteorologists was central to the use and spread of climate information throughout the wildfire management network in Arizona and New Mexico. This evaluation showed how different types of fire management information were used and how seasonal climate forecasts fit into the suite of available information (for full results, see [21]). The study revealed that the method of information delivery (e.g., person-to-person communication, an email listserv, or a forecast downloaded from the Web), as well as the relationships between people who deliver and receive that information, influence how that information is understood, used, and applied.

10.10.4   Understanding why information is not used

Evaluating the Arizona DroughtWatch drought impact reporting tool offered CLIMAS an opportunity to study a climate product that appeared to be well designed and executed, but failed to achieve the intended goal of becoming an operational component of the state of Arizona's Drought Preparedness Plan. DroughtWatch sought to provide local drought impacts data through an online drought impacts reporting system by employing volunteers around Arizona to report drought impacts in their local area.

Although it was codeveloped with intended users, participation rates were far lower than expected, severely limiting the usefulness of the tool in drought planning. The DroughtWatch evaluation clearly revealed the limits of a model built on volunteer observers and suggested the need for professional observers to provide the core data that could then be bolstered by volunteers (for full results, see [22]). The DroughtWatch evaluation also tackled a central assumption made by CLIMAS researchers over nearly the entire history of the program: a tool or product iteratively specified and built with direct participation by intended users has a high probability of success. By revealing the limits of that assumption, the DroughtWatch evaluation provided crucial insight into some of the nuances of the CLIMAS approach to transdisciplinary research.

10.10.5 Tracing the use and impact of climate information in policy making

Our final example of how evaluation activities offer insights into the mechanisms by which RISA programs build capacity comes from an external evaluation of the role of Pacific RISA in progressive adaptation planning. The external evaluation aims to examine whether appropriate stakeholders are involved, identify how progress in adaptation planning can be operationalized, examine how adaptation plans have changed over time regarding climate-change risks, and establish causal process–outcome links where possible. The wider set of motivations and obstacles to integration of information in adaptation plans is also assessed, as is Pacific RISA team functioning (e.g., quality of team communication, leadership, productivity).

More recently, external evaluation activities focused on the role of Pacific RISA in coordinating, writing, editing, and publishing a report called the Pacific Islands Regional Climate Assessment (PIRCA) [23], which was prepared in support of the Third US National Climate Assessment (NCA). Evaluation of the PIRCA process involved a multimethod, event-driven evaluation with a focus on the report's development and dissemination [24]. The purpose was to determine how broadly and in what ways the PIRCA report has reached and influenced different audiences both inside of and external to the Pacific Islands region. In addition to providing quantitative data about the PIRCA's high perceived credibility, use in decision-making and as a regional climate "consensus" document, media dissemination, and the leading role of the Pacific RISA, the external evaluation provides data on the traceable use and impact of the PIRCA in state and federal policy making, in state agency planning, and as a reference document in speeches and backing for policy initiatives by political leaders. The external evaluation also provides suggestions about how Pacific RISA can extend the reach and impact of the PIRCA process, including increasing its online availability on relevant regional websites, producing value-added report derivatives to reach more diverse audiences, and continuing to work with identified policy makers at the regional and national level to sustain the assessment process.

10.11   Summary of lessons learned from evaluating RISA regional programs

The evaluation activities by Pacific RISA and CLIMAS described above highlight several lessons regarding the opportunities afforded. First, diverse approaches are available to evaluate the effectiveness of RISA efforts to translate climate information into action in society; conceptualizations of how change occurs may be developed in multiple ways (e.g., developing an ALM, understanding stakeholder perspectives). Second, the development and use of a programmatic theory of action to describe the proposed mechanisms of change explicitly provides the opportunity to articulate and predefine measures of successful outcomes at different future times. This approach also highlights expectations (and potentially hypotheses) about how key sets of variables (e.g., context, inputs, outputs, outcomes, assumptions, external factors) work independently or in combination to determine the effectiveness of the translation process. Third, collecting qualitative and quantitative data based on specific metrics at multiple points in time encourages incremental learning and refinement of science strategies. When done well, theories of action can provide useful science policy information. For example, rather than have the policy for science (e.g., as expressed in calls for proposals) reflect simple assumptions about the value of RISA research and the outcomes of that research, a well-developed theory of action—and the evaluation data that support it—can provide a more sophisticated expression of what can be expected from the investment.

   In addition, we have learned several lessons about the challenges of evaluating RISA programs. Importantly, causality may be hard to establish because of the multiple factors changing simultaneously in a set of activities designed to address real-world climate-society challenges. Tightly controlled experimental designs are often impossible to implement and therefore unraveling the complex mechanisms by which change occurs is difficult. In addition, identifying a manageable set of valid and reliable variables to measure—without placing undue burden on stakeholders or researchers—may be challenging.

Together, the evaluation experiences of Pacific RISA and CLIMAS have offered important insights for funding priorities and decision processes at multiple levels (federal, regional, program). The lessons learned to date suggest that we can use evidence and analysis to develop a science policy that can effectively address the climate information needs of societal decision-makers.

Acknowledgments

This work was supported by the National Oceanic and Atmospheric Administration's Climate Program Office through grants NA10OAR4310216 with the Pacific RISA at the East West Center and NA07OAR4310382 and NA12OAR4310124 with the Climate Assessment for the Southwest program at the University of Arizona. The authors would like to thank the three anonymous reviewers whose comments improved this chapter substantially. They would also like to thank the book's Editorial Working Group—in particular Ryan Meyer and Adam Parris—who provided invaluable feedback and guidance throughout the development of this chapter.

References

1 National Research Council (2009) *Informing Decisions in a Changing Climate*. The National Academies Press, Washington, DC.
2 Pulwarty, R.S., Simpson, C.F. & Nierenberg, C.R. (2009) The Regional Integrated Sciences and Assessments (RISA) program: crafting effective assessments for the long haul. In: Knight, C.G. & Jäger, J. (eds), *Integrated Regional Assessment of Global Climate Change*. Cambridge University Press, Cambridge, UK, pp. 367–393.
3 National Research Council (2008) *Research and Networks for Decision Support in the NOAA Sectoral Applications Research Program*. The National Academies Press, Washington, DC.
4 Brooks, H. (1968) *The Government of Science*. MIT Press, Cambridge, Massachusetts.
5 Jahn, T., Bergmann, M. & Keil, F. (2012) Transdisciplinarity: between mainstreaming and marginalization. *Ecological Economics*, **79**, 1–10.
6 Orzag, P.R. & Holdren, J.P. (2009) *Memorandum for the Heads of Executive Departments and Agencies: Science and Technology Priorities for the FY 2011 Budget*. The White House, Washington, DC.
7 Sarewitz, D. (2011) Institutional ecology and the social outcomes of scientific research. In: Fealing, K.H., Lane, J.I., Marburger, J.H., III, & Shipp, S.S. (eds), *The Science of Science Policy: A Handbook*. Stanford University Press, Stanford, CA, pp. 337–348.
8 Bamberger, M., Rugh, J. & Mabry, L. (2012) *RealWorld Evaluation*. SAGE, Thousand Oaks, CA.
9 Teddlie, C. & Tashakkori, A. (2009) *Foundations of Mixed Methods Research: Integrating Quantitative and Qualitative Approaches in the Social and Behavioral Sciences*. SAGE, Thousand Oaks, CA.
10 Morra Imas, L.G. & Rist, R.C. (2009) *The Road to Results: Designing and Conducting Effective Development Evaluations*. World Bank Publications, Washington, DC.
11 Rossi, P.H., Lipsey, M.W. & Howard, E. (2005) *Evaluation: A Systematic Approach*. SAGE, Thousand Oaks, CA.

12 Leeuw, F.L. & Vaessen, J. (2009) *Impact Evaluations and Development: NONIE Guidance on Impact Evaluation*. The Network of Networks on Impact Evaluation, Washington, DC.

13 Funnell, S.C. & Rogers, P.J. (2011) *Purposeful Program Theory: Effective Use of Theories of Change and Logic Models*. Wiley, San Francisco.

14 Stufflebeam, D. (2001) Evaluation models. *New Directions for Evaluation*, **2001** (**89**), 7–98.

15 National Research Council (2005) *Thinking Strategically: The Appropriate Use of Metrics for the Climate Change Science Program*, pp. 150. National Academies Press, Washington, DC. http://www.nap.edu/catalog/11292.html.

16 NOAA Climate Program Office (2014) *About the Regional Integrated Science and Assessments Program*. URL http://cpo.noaa.gov/ClimatePrograms/ClimateandSocietalInteractions/RISAProgram/AboutRISA.aspx [accessed on 8 February 2014].

17 Wallsgrove, R., Penn, D.. (2012) *Water Resources and Climate Change Adaptation in Hawai'i: Adaptive Tools in the Current Law and Policy Framework*. Center for Island Climate Adaptation and Policy, Honolulu, HI. URL http://icap.seagrant.soest.hawaii.edu/icap-publications.

18 Michaels, S. (2009) Matching knowledge brokering strategies to environmental policy problems and settings. *Environmental Science & Policy*, **12** (**7**), 994–1011. doi:10.1016/j.envsci.2009.05.002.

19 Cash, D.W., Clark, W.C., Alcock, F. *et al.* (2003) Knowledge systems for sustainable development. *Proceedings of National Academy of Sciences*, **100** (**14**), 8086–8091.

20 Guido, Z., Hill, D., Crimmins, M. & Ferguson, D. (2013) Informing decisions with a climate synthesis product: implications for regional climate services. *Weather, Climate & Society*, **5** (**1**), 83–92.

21 Owen, G., McLeod, J.D., Kolden, C.D. *et al.* (2012) Wildfire management and forecasting fire potential: the roles of climate information and social networks in the southwest United States. *Weather, Climate & Society*, **4**, 90–102.

22 Meadow, A.M., Crimmins, M.A. & Ferguson, D.B. (2013) Field of dreams or dream team? Assessing two models for drought impact reporting in the Semiarid Southwest. *Bulletin of the American Meteorological Society*, **94** (**10**), 1507–1517.

23 Keener, V.W., Marra, J.J., Finucane, M.L. *et al.* (eds) (2013) *Climate Change and Pacific Islands: Indicators and Impacts: Report for the 2012 Pacific Islands Regional Climate Assessment*. Island Press, Washington, DC.

24 Moser, SC. (2013) *PIRCA Evaluation: Development, Delivery, and Traceable Impacts – With Particular Emphasis on the Contributions of the Pacific RISA*. Independent Evaluation Report. Susanne Moser Research and Consulting, Santa Cruz, CA. URL www.PacificRISA.org [accessed on 7 September 2015].

CHAPTER 11

Navigating scales of knowledge and decision-making in the Intermountain West: implications for science policy

Eric S. Gordon[1], Lisa Dilling[1,2], Elizabeth McNie[1] and Andrea J. Ray[1,3]

[1] Western Water Assessment, Cooperative Institute for Research in Environmental Sciences, University of Colorado Boulder, 216 UCB, Boulder, CO 80309, USA

[2] Environmental Studies Program and Center for Science and Technology Policy Research, Cooperative Institute for Research in Environmental Sciences, University of Colorado Boulder, 4001 Discovery Drive, Boulder, CO 80309-0397, USA

[3] Physical Sciences Division, NOAA Earth System Research Laboratory, 325 Broadway, R/PSD1, Boulder, CO 80305, USA

11.1 Introduction

Defined simply as the "spatial, temporal, quantitative, or analytical dimensions used to measure and study any phenomenon," [1] scale is a key analytical and explanatory attribute of the human–environment system [2]. Considerations of scale are fundamental to investigating and understanding how to support research, outreach, and engagement with decision-makers who need useful information to expand policy alternatives, clarify choices, and otherwise improve policy outcomes [3]. Unfortunately, navigating across multiple scales of research and decision-making is a difficult task for many traditional research entities, and failure to actively manage multiscalar challenges can lead to the production of information that is not useful for decision support.

In this chapter, we illustrate how the Western Water Assessment (WWA) has identified and addressed problems of scale in order to support climate-sensitive decision-making by water resource managers in the Intermountain West. One of the oldest Regional Integrated Sciences and Assessments (RISA) programs, WWA began in 1998 as an initiative among scientists

Climate in Context: Science and Society Partnering for Adaptation, First Edition.
Edited by Adam S. Parris, Gregg M. Garfin, Kirstin Dow, Ryan Meyer, and Sarah L. Close.
© 2016 John Wiley & Sons, Ltd. Published 2016 by John Wiley & Sons, Ltd.

at the University of Colorado's Cooperative Institute for Research in Environmental Sciences and the National Oceanic and Atmospheric Administration's (NOAA) Climate Diagnostics Center. WWA has since evolved into a full-fledged RISA working in Colorado, Wyoming, and Utah, a region characterized largely by semi-arid grasslands, high-elevation mountain ranges, and desert basins. WWA's mission is to "conduct innovative research and engagement aimed at effectively and efficiently incorporating knowledge into decision making in order to advance the ability of regional and national entities to manage climate impacts." Given the water scarcity in the region, WWA's primary focus has been on the water sector.

Much has been written about the concept of scale and how it is used across the natural and social sciences, especially when examining how science can better inform efforts to govern in the face of environmental change [1,4,5]. Scholars have focused on improving the precision with which we understand and use scalar concepts (e.g., temporal, spatial, jurisdictional, institutional; see [6]) and levels within scales (e.g., daily/seasonal/annual, or local/state/national.) The concept of "mismatches" or lack of "fit" among levels of spatial scale such as watersheds, levels of jurisdictional scale of governance such as state boundaries, and levels of institutional scale such as regulatory rules is well recognized (e.g., [1,6]). Challenges have also been identified in connecting the spatial validity of data for a given level of application and the needs of decision-makers at particular jurisdictional levels (e.g., applying global models to local resource decisions) [4,7]. In this chapter, we draw on these concepts but do not attempt to capture the scale discussion in its entirety. Instead, we focus on examples from WWA's experience that illustrate how scale and level considerations influence the production and application of knowledge in decision-making. With particular reference to knowledge, we examine who produces the knowledge, at what organizational level, and with what spatial characteristics relevant to the level of jurisdiction and decision-maker.

As a RISA, WWA provides an ideal opportunity to ask these questions and observe the challenges of knowledge production and use across spatial levels and jurisdictional boundaries. The geographic and jurisdictional levels of operation of RISAs are deliberately cross-cutting and loosely defined, giving them the freedom to seek out knowledge and knowledge producers from a variety of spatial and institutional levels in order to connect with decision-makers at multiple levels within a region. Rather than acting as a constraint, a regional focus provides scale-dependent advantages, allowing RISAs to experiment with new sources of knowledge, connecting decision-makers with previously unfamiliar knowledge providers or expanding the scope of knowledge provision to new types of decision spaces. The guiding principle for RISA work is to begin with the decision context of particular stakeholders in a region and to let that context shape the sources and types of knowledge that are brought to bear on decision-making. Moreover, RISAs have been able to interact with actors at multiple scales from both sides of the science–policy boundary because they are free from the strictures of official government-provided climate services and because they are still able to maintain longstanding relationships with researchers and decision-makers. This also enables RISAs like WWA to be nimble, flexible, and experimental in order to adapt to changing scientific and policy windows of opportunity.

Intertwined with considerations of scale are additional research findings demonstrating that decision-makers are more likely to use scientific information if it is considered to be salient, credible, and legitimate [8,9]. *Salient* information is inherently sensitive to context and relevant to the particular spatial, temporal, jurisdictional, and institutional scales of a problem. For example, climate information developed and presented at a national level of spatial scale can be so broad that it may not apply well to local-level decisions. Whether or not information is *credible* relates to users' perceptions of accuracy and quality of the data or perceptions about the standing of the knowledge producer. Users also view information as *legitimate* when they believe it was produced free from political persuasion or bias and when their needs and concerns were somehow incorporated into that production process. Finally, useful information takes on a procedural dimension in which producers and users of information engage in a dialogue aimed at shaping research agendas based on the context and scale-dependent needs of decision-makers [10,11]. In our experience, producing usable and salient information for decision support means that we must consider how to navigate multiple spatial levels and dimensions, but also ensure that WWA produces high-quality information and is viewed by decision-makers as a trustworthy source of knowledge.

In this chapter, we show how WWA has engaged in this process of "navigation" to make climate knowledge, produced at multiple spatial and temporal levels, useful in supporting public decision-making at multiple jurisdictional levels. The WWA experience demonstrates some of the challenges facing this process, such as identifying information relevant to a particular decision context and providers capable of creating and/or delivering such information; navigating scales to provide salient, credible, and legitimate knowledge for particular contexts; and addressing the unfamiliarity or lack of trust between relevant knowledge providers and decision-makers who each have relevant contributions to the problem.

We draw from existing literature and theories to demonstrate these scale challenges and then share five case studies from WWA's work in order to explore how strategies of convening, translating, collaborating, and mediating, commonly used in boundary work, can help navigate scale-dependent problems. We conclude with a discussion about science policies needed to support climate research aimed at decision support. Specifically, we argue that science policies need to be changed in order to ensure that grants are of adequate duration to support building and sustaining relationships, that research organizations are nimble and flexible enough to respond to emerging opportunities, and that boundary work is adequately evaluated and incentivized by funding agencies, universities, and other research facilities.

11.2 Doing boundary work to overcome scale challenges

Decision-makers need climate information that is salient, credible, and legitimate to help inform decisions related to climate adaptation. When called upon to respond to specific information needs, however, traditional science producers often provide information developed to answer scientific questions rather than information intended to be useful in specific decision contexts. Moreover, science producers are often unaware of common scale-related problems (see Table 11.1) that challenge the effective integration of climate information into decision-making [3,4]. Producing usable information requires that the work of scientists and the needs of decision-makers be reconciled through increased interaction between both groups. There are pitfalls in this process, however—too much involvement of science in policy decisions can lead to the politicization of science and reduced credibility, while insufficient interaction may result in the production of more information without improving usefulness.

Striking the right balance, therefore, requires "boundary work" to actively manage the boundary between science and policy [12]. Boundary work involves four key strategies: *convening* different actors to produce useful information, *translating* information to actors on both sides of the science–policy divide, providing opportunities for *collaboration* in research or product development, and *mediating* problem framing and conflict among actors [12,13]. This work is carried out by boundary organizations [12] whose function is to straddle the boundary between science and policy and strive to increase linkages between science and policy while simultaneously working to preserve scientific credibility. When done effectively, boundary work can increase the usefulness of climate information in scale-dependent contexts [4]. WWA and other RISA programs have often been described as "boundary organizations," meaning that they are entities specializing

Table 11.1 Major scale discordance problems affecting societal responses to climate-related challenges.

Scale challenge	Definition
Institutional fit problem	Mismatches between the spatial scale of environmental phenomena and the geographic scale of political entities (e.g., counties, cities, or states).
Scale discordance problem	Mismatches occurring when available scientific information does not reflect the unique context of the environmental conditions and/or the geographic scale of decision-making; often arises in trying to apply general research findings to specific contexts or when efforts to assess climate impacts are conducted at scales not relevant to decision-making.
Insufficient attention to cross-scale linkages	Mismatches related to an over-reliance on scientific causal explanations at one particular scale at the expense of identifying other causal relationships at different scales; often arises when scientists undertake research aimed at a single scale of decision-making and miss cross-scale interactions that affect decision contexts.

Source: from [4].

in boundary work [14] and thus provide ideal opportunities to examine the value of such organizations in navigating scales to provide climate information for decision support.

This chapter uses case studies of specific WWA processes or outputs to illustrate the efforts involved in such boundary work. The examples provided in the following text all demonstrate the multiple scales and levels of scales at issue in a given context, one or more climate-related scale challenges, the use of boundary work strategies in meeting scale-related challenges, and lessons learned from addressing relevant scale challenges. Figure 11.1 provides a visual demonstration of how each of the cases crossed multiple levels of knowledge production and decision-making.

11.2.1 Reservoir management and endangered species workshop

In 1995, the NOAA Climate Prediction Center (CPC) began issuing a new generation of seasonal predictions based on the El Niño-Southern Oscillation (ENSO) climate phenomenon. WWA took advantage of these predictions for its first major event, a 1999 workshop that brought together WWA partners, potential stakeholders, WWA researchers, and water managers. The assembled group identified a potential area of collaboration to explore

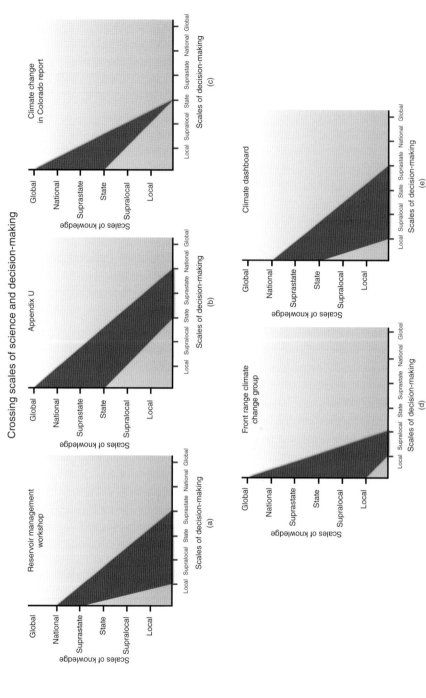

Figure 11.1 (a)–(e) Individual examples of how WWA's work crossed multiple scales. On the y-axis are scales of knowledge shown at the level of the climate knowledge producer, while the x-axis shows scales of decision-making. "Supralocal" refers to decision-making or knowledge production entities that operate at a scale greater than an individual municipality but less than an entire state, while "suprastate" refers to entities operating at scales larger than an individual state but less than national.

the benefits of applying these seasonal predictions to water supplies on the Upper Colorado River in Colorado. That fall, WWA and two important stakeholders in reservoir management, the Colorado River Water Conservation District (CRWCD) and Denver Water, co-convened the "Workshop on Weather and Climate, Reservoir Operations, and Endangered Fish in the Upper Colorado River Basin." This workshop was designed to provide an opportunity for NOAA, university scientists, and other climate information providers to discuss available ENSO-related research and products. In turn, water resources decision-makers informed researchers about reservoir operations, the challenge of meeting flow requirements for endangered fish, and whether current climate information could be used in planning. Overall, the workshop attempted to bring the understanding of ENSO impacts and seasonal predictions from NOAA's national seasonal climate outlooks to the context of decision-making for water resources in the Colorado River headwaters region. (Note: This workshop is discussed further in Chapter 2.)

The institutional and jurisdictional levels at issue here illustrate the complexity of the decision-making context (Figure 11.1(a)). Water management stakeholders involved in the workshop included a federal agency (the U.S. Bureau of Reclamation, hereafter "Reclamation"), two supralocal entities (the CRWCD and the Northern Colorado Water Conservancy District), and a local-level water utility (Denver Water) whose operations span two water basins, with its service territory located in the South Platte Basin but with significant additional supplies and storage capacity in the Colorado River headwaters area. The US Fish and Wildlife Service was also present because of their mandate to restore habitat for endangered fish downstream on the Colorado River. This mandate was challenging managers at multiple spatial and institutional scales to come up with a strategy to provide water for fish under the federal Endangered Species Act. Thus, water management in this basin was at a point of criticality and at a nexus of local, state, suprastate, and federal levels of jurisdictional and institutional scales. Knowledge, meanwhile, was being produced at a regionally relevant level by national research institutions. Throughout this process, WWA faced the dual task of helping users see that knowledge as salient, credible, and legitimate, while also providing feedback to NOAA on improving its usability. Over the next several years, WWA funded studies to both understand the user context (described in [15] and Chapter 2) and to facilitate the use of National Weather Service 8–14 day outlooks and NOAA CPC seasonal outlooks in river management [16]. WWA also regularly engaged with both the CPC and the NOAA Colorado Basin River Forecast Center to improve the usability of their products.

Crossing multiple levels along these multiple scales raised a number of challenges. The first arose from the fact that seasonal predictions of temperature and precipitation related to ENSO were not being incorporated into NOAA's streamflow forecasts for the Colorado River Basin. WWA scientists also were concerned about the level of skill involved in the seasonal predictions and about ensuring that the managers understood that skill. Thus, the workshop intended to discover if more skillful forecasts might have a place in reservoir management and what temporal and spatial levels of forecasts would be most salient. The complex nature of reservoir operations conducted by a variety of entities operating under different sets of rules and practical constraints led to a cross-scale concern; without an understanding of the myriad interconnections among the entities involved, including who influenced decisions, efforts to bring climate information to bear would be of limited value.

To address these problems, WWA used multiple boundary work strategies. First, it served a convening function, providing an opportunity for knowledge producers and decision-makers across scales and levels to collaboratively explore the potential uses of climate knowledge such as seasonal climate and streamflow predictions, a process that was ongoing for several years. This procedural dimension—engaging over time—is described in Lemos and Morehouse [17] as part of the "iterativity" required to produce usable information. Scale issues were also addressed through translation by providing explanations of the skill of the prediction products in a form accessible to water managers who were technically savvy professionals, not trained in climate science. WWA also sought to provide climate information that might expand the range of choices and options for reservoir managers, potentially expanding the range of possible options for meeting ESA requirements for environmental flows. We consider this to be a type of *mediating* function—providing information that might reduce the potential for conflict among water managers.

Internally, WWA saw the workshop to be an early success because many of the water management entities represented became active partners in WWA's future work, relationships that have been sustained through to the present. Thus, WWA's use of boundary work strategies allowed it to progress from initial relationship building through the workshop to the collaborative production of research and then to a position as an expert voice on interpreting information regarding climate change.

11.2.2   Appendix U

The Colorado River provides critical water supplies to seven states under a complicated set of rules defined by the Colorado River Compact. Prior to

2005, no plan existed for managing the river's two major reservoirs (Lakes Powell and Mead) during a drought of magnitude sufficient to prompt the states in the Lower Basin to "call" for water from the Upper Basin. In light of an unprecedented drought during the 2000s, however, Reclamation began a process of developing new operating criteria for the two reservoirs, ultimately resulting in the "Colorado River Interim Guidelines for Lower Basin Shortages and Coordinated Operations of Lake Powell and Lake Mead" [18]. WWA, which had a working relationship with Reclamation's Upper and Lower Colorado River regional offices since the late 1990s, provided leadership and guidance as part of the Climate Change Technical Work Group (TWG) that produced "Appendix U," a detailed assessment of the impacts of climate change on future flows in the Colorado River. Appendix U was included as part of the final Environmental Impact Statement (EIS) for the new drought management criteria.

As shown in Figure 11.1(b), the development of Appendix U involved a wide variety of knowledge and decision-making levels. The primary level of decision-making involved a federal agency (Reclamation) that was engaged in a formal environmental impact analysis as required by law for a decision of such magnitude. However, the geographic scale of the EIS itself was suprastate—involving the Colorado River Basin—and strongly tied to decision making and planning at the state level, given that seven states are party to the Colorado River Compact. On the other hand, knowledge used in the report was generally generated at higher levels, such as global climate models and the suprastate-scale Colorado River Simulation System model used by the TWG to translate climate projections into future hydrology scenarios.

For WWA, the major scale challenge in this case was an institutional fit problem—the climate knowledge needed for this effort was scattered among many science institutions and produced at a variety of levels, not all of which were directly relevant to assessing projections of future flow in the Colorado River. Research institutions rarely possess the in-house capacity to quickly adapt their knowledge to the spatial levels needed for the decision context. In addition, this case also illustrates a temporal institutional fit problem—Reclamation needed climate information provided in significantly less time than would normally be possible under traditional research time frames.

To address these problems, WWA worked with Reclamation and other RISAs to convene the TWG, which included experts in meteorology, climate, and hydrology from WWA, other RISAs, NOAA, the U.S. Geological Survey, and a consulting firm. Once convened, the TWG collaborated with Reclamation to produce relevant information by convening multiple research groups to frame and synthesize appropriate climate knowledge. The TWG was able to bring to bear science on two temporal levels not previously reflected in Reclamation's analyses—paleoclimate information to reflect conditions out of the range of the historic record and projections of future climate change. Meeting the temporal challenge (the work needed to be done in less than a year to meet the needs of the EIS process) required WWA to harness its scale-dependent advantages, helping the TWG synthesize information and produce new knowledge in a time-frame rarely seen in traditional research. This effort also demonstrated the value of translation as a strategy for incorporating climate science into a document relevant to the work of water resources engineers.

The funding flexibility provided by the RISA program gives WWA the ability to allocate resources based on emerging problems. In this case, WWA was able to successfully facilitate the work of the TWG—including funding its scientists to participate—in developing climate information to meet a "policy window" [19] that would be too short for a typical research grant to respond. Collaborative efforts undertaken for Appendix U also led to the development of a separate multi-stakeholder project that evaluated and synthesized existing research on future warming impacts and projections for overall flow in the Colorado River [20]. Through these efforts, the group was able to integrate the research developed at multiple spatial and institutional levels to provide information that was more useful and directed to the interests of Colorado River stakeholders than any single effort could have independently accomplished.

11.2.3 Climate Change in Colorado

In 2007, Colorado Governor Bill Ritter issued a Climate Change Action Plan for the state that called for state agencies to "prepare the state to adapt to those climate changes that cannot be avoided" [21]. After that plan was issued, the Colorado Water Conservation Board (CWCB, the state agency charged with long-term planning and management of water resources) commissioned WWA to help compile relevant climate science into a report oriented toward the water resources sector. WWA had been interacting with the CWCB since the 1999 workshop described earlier, so the idea for such a project came out of ongoing discussions. The resulting report, *Climate Change in Colorado* [22], was collaboratively produced by WWA and the CWCB, with inputs from other water management entities across the state. It provided a synthesis of the existing climate observations and projections as well as research on understanding potential changes to surface water supplies under a warming climate.

As shown in Figure 11.1(c), a variety of spatial levels of knowledge production were brought to bear on the development of this report, which was in turn oriented toward decision-making at a specific set of jurisdictional levels. The report authors drew from global-level research (e.g., global climate model projections), suprastate-level research such as climate change impacts to flow in the Colorado River, and state-level information such as observed records of the state's climate. CWCB wanted a product that would inform state-level decision-making, although it was also useful for water management within larger entities like Reclamation and smaller entities like Denver Water.

Climate Change in Colorado primarily sought to address a scale discordance problem between knowledge produced at global spatial and institutional levels and decision contexts at the state and local jurisdictional levels. At the time, a number of scientific assessments of climate change impacts had been produced—notably the IPCC Fourth Assessment Report (AR4) [23]; policymakers in Colorado were interested in understanding more precisely how this information was relevant to water management in the state. AR4 was produced at a global scale, providing at best continental-scale analyses and information oriented toward national and international decision-making. CWCB needed information oriented toward state-level decision-making, which prior to creation of *Climate Change in Colorado* had either not existed or not been readily available. Thus, the challenge in developing the report was to translate knowledge produced at global, suprastate, and state scales into a form sufficient for supporting state-level decision-making.

To solve this scale discordance problem, WWA relied primarily on translation and convening strategies. The translation strategy was employed not simply by synthesizing global and federally produced climate information for the state and supralocal levels, but also by crafting a report that was relatively free of scientific jargon and was comprehensible to decision-makers. In addition, WWA and CWCB both provided funding and leadership that enabled the convening of scientists from multiple entities, including NOAA's Physical Sciences Division, the University of Colorado Boulder, and Colorado State University. A number of water management agencies (listed as contributors in that report) collaborated by participating in meetings to study the scope and design and review the report. WWA was able to harness its ability to operate at the right jurisdictional and spatial levels while coordinating among entities with overlapping. The report underwent a rigorous review process, which included reviewers from both sides of the science–policy divide, giving it legitimacy in the scientific and decision-making worlds. The result was a co-branded report that provided useful, decision-relevant summaries of climate observations and projections for the state. As a measure of its usability and relevance, the report has been used repeatedly by the CWCB to demonstrate the need to account for climate change in future water supply planning. In addition, the CWCB requested an updated version of the report subsequent to the release of new global climate modeling efforts [24].

11.2.4 Front Range Climate Change Group

As described in [25], municipal and industrial water utilities (M&Is) along Colorado's Front Range urban corridor evolved from only considering seasonal climate information in operations to actively examining potential climate change impacts on supplies and integrating those impacts into planning efforts. Building on its previous collaborations with Front Range water providers, in 2006 WWA provided a workshop on climate change and potential impacts on Front Range water supplies, followed by a 2008 workshop on climate modeling. The utilities interested in the process then began an affiliation known as the Front Range Climate Change Group (FRCCG). In collaboration with WWA, the FRCCG worked with a consulting group to conduct a formal study (called "The Joint Front Range Climate Change Vulnerability Study") to examine the vulnerability of their shared water-supply resources. In 2013, WWA began working with the FRCCG to provide a series of continuing educational workshops on climate-related issues.

Figure 11.1(d) demonstrates that WWA's work with the FRCCG is mostly relevant to local decision-making, particularly planning and operations at M&Is. However, the FRCCG is also a supralocal entity—in other words, a group of local entities that consider the effects on resources they share at spatial and jurisdictional levels greater than their own individual service boundaries. As is common with many of these efforts, climate knowledge generated at multiple spatial levels was brought to bear on the FRCCG—from global climate model output to supralocal streamflow data from the Colorado Decision Support System to localized knowledge about vulnerabilities to a watershed or particular provider.

Similarly to the *Climate Change in Colorado* example, the FRCCG's activities were subject to institutional fit and scale discordance problems. The M&Is engaged in a joint effort largely because they use supply sources located well beyond the boundaries where they deliver water. Moreover, some of these utilities were looking at the same water sources to provide additional supplies as buffers against climate change-related decreases in supply. To provide information that would help making more informed decisions in the face of these problems, the group needed climate science produced at multiple levels in order to develop information relevant to decision-making in the Front Range context.

More importantly, perhaps, an institutional fit problem arose from the nature of the entities collaborating on the development of the study. Many science entities that develop climate-projection information are accustomed to working on longer time frames through a traditional grant-research model, whereas the M&Is are accustomed to rapid scoping and development of projects, primarily working through a consultant-client model. To deal with that hurdle, the FRCCG used a consulting company as the technical authors of the study but enlisted WWA and other scientific experts in the scoping and review of the project. WWA helped move this process along using convening and collaborating strategies, but it is important to note that FRCCG members themselves led the effort, not WWA. In its continuing role in facilitating climate literacy improvements through a translation strategy, however, WWA has worked on an ongoing basis to provide further workshops and other materials to help the group stay abreast of emerging climate-related issues.

WWA's interactions with the FRCCG relied on boundary work to help facilitate the process of bringing together experts and stakeholders to develop shared knowledge about climate change in a specific context. Particularly critical to this success was the ability of WWA to participate in rapidly developed efforts aimed at understanding the utilities' common vulnerabilities. In this instance, although the FRCCG members were more accustomed to working directly with consulting companies to produce reports, WWA was key enough to the overall effort that one of the members of the FRCCG referred to WWA as "the most effective and beneficial model for meeting our education and assessment needs" [26].

11.2.5 Intermountain West Climate Dashboard

Since 2005, WWA has produced a climate summary oriented toward water and other resource decision-makers in its three-state region. Initially, WWA sent out a monthly "Intermountain West Climate Summary" (IWCS; modeled after the "Southwest Climate Outlook" created by the Climate Assessment for the Southwest RISA) providing graphics and recent climate conditions and seasonal forecasts accompanied by a narrative explanatory text. IWCS issues frequently included articles aimed at improving readers' climate literacy or introducing new WWA work. The IWCS primarily served as a single point of reference for multiple sources of information developed at a variety of spatial and institutional levels of knowledge production. In 2012, WWA replaced the IWCS with a dashboard-style website, the Intermountain West Climate Dashboard (IWCD), which provides real-time versions of the same information using more advanced web technology along with short explanatory text updates.

The IWCD is perhaps the broadest cross-scale interaction WWA engages in (see Figure 11.1(e)), largely because of the variety of producers and the number of entities interested in similar information. The primary users of the dashboard (based on a list of approximately 500 email subscribers) include entities at virtually all jurisdictional levels, from local (e.g., individual M&Is) to suprastate/national (e.g., Reclamation). Information displayed on the website includes products produced primarily at the national (e.g., seasonal forecasts from NOAA's CPC) and suprastate (e.g., precipitation maps from the High Plains Regional Climate Center) levels by climate service institutions that generate much of the climate information available in the United States.

Alone, these products are often not matched well to the information needs of many decision-makers. Collectively, however, having the IWCD as a single point of reference allows decision-makers to assimilate information from a variety of scales for their purposes. The IWCD tackles the problem of integrating information from multiple scales by gathering information from diverse sources through a web-based tool, resulting in the production of a useful cross-scale product that cannot otherwise be created by individual climate information producers.

The IWCD (and its predecessor, the IWCS) has successfully played two roles in crossing scales—first, by translating a variety of climate information products available from diverse and often uncoordinated sources into formats more easily used by subscribers; and second, as a means to draw stakeholders' attention to WWA's work by creating a subscriber list that receives other updates on WWA's work. A 2008 survey done by WWA demonstrated that a core group of IWCS subscribers found the summary to be an efficient way to access important climate information. Although no similar survey has been conducted for the IWCD, initial feedback from specific users indicated that the new format provided the same information in a more efficient manner. Moreover, imitation may be a measure of success—the IWCD has already been replicated by other climate service organizations, including the Colorado Climate Center.

11.3 Implications for science policy

Our aim in this chapter has been to demonstrate how WWA's use of boundary work and its structural flexibility helped to bridge climate knowledge production and decision-making across multiple scales. These efforts stand in stark contrast to the bulk of climate-related research, which is aimed at expanding our general understanding of climate phenomena by testing hypotheses and informing various theories. Most climate research also generally focuses on large-scale processes that are rarely relevant to decision-makers without extensive translation.

Transforming climate research, along with other climate information not produced in the context of user needs, into forms useful for decision-makers requires boundary work, particularly when attempting to navigate the kind of scale-related challenges we have identified in this chapter. Unfortunately, existing science policies and basic research structures often do not support boundary work and even create disincentives for doing it. Producing usable information will thus require shifting science policy and funding to better support boundary work in research organizations. Specifically, science funding entities should consider offering more grants that are of adequate duration to support building and sustaining relationships. Research organizations should aim to be nimble and flexible in order to respond to emerging problems, and boundary work should be adequately evaluated and incentivized by funding agencies, universities, and research facilities.

These shifts in science policy are critical for orienting climate knowledge production to decision support, as demonstrated not only by WWA's experience but also by the lessons learned in a variety of other contexts (e.g., [9, 10]). For example, convening diverse groups of decision-makers and researchers is predicated on relationships based on mutual trust and respect, which takes resources and time to accomplish. The ability to convene decision-makers and researchers enables boundary organizations like WWA to better assess users' information needs, tailor research to respond to those needs, understand the context in which climate science will be used, connect different actors to each other, and efficiently use resources to conduct, communicate, and translate research. Convening researchers was particularly important in WWA's work on the Climate Change in Colorado Report, as well as with the FRCCG. Moreover, time spent building relationships in one instance—such as during the Reservoir Management and Endangered Species Workshop—yielded future benefits when established relationships were leveraged in emerging projects. WWA's experience suggests that this sort of convening may take years of cultivating relationships before they can be leveraged for successful outcomes. Typical grant funding cycles, however, last between 1 and 5 years, which may not be enough time to build relationships and carry out boundary work to produce usable information. Science policies aimed at producing usable information for decision support should therefore be structured to allow more time to build relationships, particularly in the early stages of the funding cycle when convening work is just getting underway. Providing funding support from 5 to 10 years may be more appropriate for boundary work especially in contexts where societal problems are culturally, politically, or scientifically complicated.

Developing long-term relationships with stakeholders is also critical to enabling boundary organizations to be ready to respond to emerging problems in order to produce useful information. WWA's experience with Appendix U demonstrated its ability to take advantage of a policy window by convening diverse groups of researchers, collaborating on research, and producing a report in a single year. This was due to the program's organizational design, enabling it to reallocate human and financial capital to take advantage of emerging opportunities. WWA has the discretion to apply a sizeable portion of its budget on projects that it does not have to identify in its initial grant proposal, enabling it to be flexible and nimble in shaping its research agendas. In contrast, most federally-funded researchers are required to explain their entire budget in detail, locking in research agendas before they have received any funding or adequately assessed users' information needs. RISA programs are given more latitude to make decisions about how to spend their money during their funding period, allowing them to be more entrepreneurial, thereby shaping their research to best respond to emerging opportunities. Other research efforts aimed at producing usable information for decision support could learn from this model by providing more discretion in allocating funds and shaping research agendas during funding periods.

Reconsidering traditional research incentives could also help support efforts to make climate information usable. Boundary work such as convening, translating, collaborating, and mediating is rarely rewarded in the basic research community, in universities, or at federal research facilities; yet it is critically important in producing usable climate information for decision support. One of WWA's most widely used products, the IWCD, is largely a translational effort. The Reservoir Management and Endangered Species Workshop and the Climate Change in Colorado Report both depended on strong convening and mediation efforts by WWA. Such boundary work, however, results in outputs and outcomes that may be ambiguous and difficult to quantify, especially when compared to the standard peer-reviewed publications used to evaluate most research. Doing this type of work is not only difficult to evaluate, but disincentives at most research institutions often make researchers reluctant to participate. For example, tenure, retention, and promotion decisions at universities are largely based on the number and quality of one's peer-reviewed publications, forcing early career researchers to weigh the possibility that their efforts at boundary work may not only be missed in the evaluation process, but may also be seen as an unnecessary distraction to doing research that results in peer-reviewed publications. Science policies need to consider the development of incentives and evaluation that properly reward boundary work [9]. Research grants need to be explicit about supporting boundary work and ensure that such work will be considered in evaluating the success of the research program. Moreover, universities and research facilities that are interested in supporting use-inspired research need to expand their criteria of what constitutes worthwhile activities when evaluating tenure, retention, and promotion.

11.4 Conclusions

Multiple challenges face research organizations that work across the boundary of science and policy and strive to produce and deliver usable climate information that expands alternatives, clarifies choices, and enhances capacity to adapt to a changing climate. Over the past decade and more, the loose definition of being a "regional" entity has been used for WWA to overcome many of these challenges. Other scale-dependent advantages include being able to draw from a variety of knowledge sources produced at multiple institutional and jurisdictional levels while also interacting credibly in decision contexts at multiple levels. WWA's role as an autonomous research organization allows it to convene researchers, decision-makers, and other stakeholders from national to local levels and across sectors. This in turn facilitates building collaborative relationships with knowledge producers from diverse disciplinary, geographic, and agency affiliations, translating complex scientific information into salient and useful formats, and providing information about climate risk in appropriate policy contexts. While WWA rarely engages in direct mediation efforts among conflicting parties, its ability to provide information that expands policy options has helped ameliorate some water resources conflicts in the Intermountain West.

Ultimately, however, the ability of WWA and other RISAs to succeed in their efforts to navigate across scales and provide climate information relevant to decision support is hampered by a variety of science policy constraints that have been just described. Despite the U.S. Global Change Research Program's new strategic plan calling for more use-inspired research, any substantive changes are yet to be seen, which would support more RISA-like research. Science policies have been slow to support the growing need for usable climate information, echoing earlier criticisms made about the program [11,27–29]. Conducting basic research aimed at contributing to our fundamental knowledge about climate change will always play a critical role in climate science. Use-inspired research, however, does not drive out basic research, so it is not an "either/or" dilemma. As we demonstrated in our cases (and in other research; e.g., [14]), directing research toward producing usable information for decision support can achieve multiple goals simultaneously: it can address scale-dependent challenges, produce information that is salient, credible and legitimate, and contribute fundamental knowledge about climate science.

References

1 Gibson, C.C., Ostrom, E. & Ahn, T.K. (2000) The concept of scale and the human dimensions of global change: a survey. *Ecological Economics*, **32** (**2**), 217–239.

2 Clark, W.C. (1985) Scales of climate impacts. *Climatic Change*, **7**, 5–27.

3 Sarewitz, D. & Pielke, R. Jr., (2007) The neglected heart of science policy: reconciling supply and demand for science. *Environmental Science and Policy*, **10** (**1**), 5–16.

4 Cash, D.W. & Moser, S.C. (2000) Linking global and local scales: designing dynamic assessment and management processes. *Global Environmental Change*, **10**, 109–120.

5 Cash, D.W., Adger, W.N., Berkes, F. *et al.* (2006a) Scale and cross-scale dynamics: governance and information in a multilevel world. *Ecology and Society*, **11** (**2**), 8.

6 Young, O.R. (2003) Environmental governance: the role of institutions in causing and confronting environmental problems. *International Environmental Agreements*, **3**, 377–393. doi:10.1023/B:INEA.0000005802.86439.39.

7 Kates, R.W., Clark, W.C., Corell, R. *et al.* (2001) Sustainability science. *Science*, **292** (**5517**), 641–642.

8 Cash, D.W., Clark, W.C., Alcock, F. *et al.* (2003) Knowledge systems for sustainable development. *Proceedings of the National Academy of Sciences*, **100** (**14**), 8086–8091.

9 Dilling, L. & Lemos, M.C. (2011) Creating usable science: opportunities and constraints for climate knowledge use and their implications for science policy. *Global Environmental Change*, **21**, 680–689.

10 Morss, R.E., Wilhelmi, O.V., Downton, M.W. *et al.* (2005) Flood risk, uncertainty, and scientific information for decision making: lessons from an interdisciplinary project. *Bulletin of the American Meteorological Society*, **86**, 1593.

11 McNie, E. (2007) Reconciling the supply of scientific information with user demands: an analysis of the problem and review of the literature. *Environmental Science and Policy*, **10** (**1**), 17–38.

12 Guston, D.H. (2001) Boundary organizations in environmental policy and science: an introduction. *Science, Technology, and Human Values*, **26** (**4**), 399–408.

13 Cash, D.W., Borck, J. & Patt, A. (2006) Countering the loading-dock approach to linking science and decision making. *Science, Technology and Human Values*, **31**, 30.

14 McNie, E. (2008) *Co-Producing Useful Climate Science for Policy: Lessons from the RISA Program*. PhD Dissertation, University of Colorado at Boulder, Boulder, CO.

15 Ray, A. (2004) *Linking Climate to Multi-Purpose Reservoir Management: Adaptive Capacity and Needs for Climate Information in the Gunnison Basin, Colorado*. Dissertation, Dept. of Geography, University of Colorado, Boulder, CO, 328 pp.

16 Clark, M.L., Hay, G., McCabe, G. *et al.* (2003) Use of weather and climate information in forecasting water supply in the western United States. In: Lewis, W.J. (ed), *Water and Climate in the Western United States*. University Press of Colorado, Boulder, CO.

17 Lemos, M.C. & Morehouse, B.J. (2005) The co-production of science and policy in integrated climate assessments. *Global Environmental Change*, **15**, 57–68.

18 US Department of the Interior, Bureau of Reclamation (2007) *Final Environmental Impact Statement: Colorado River Interim Guidelines for Lower Basin Shortages and Coordinated Operations for Lake Powell and Lake Mead*.

19 Kingdon, J.W., Thurber, J.A. (1984). *Agendas, Alternatives, and Public Policies* (Vol. **45**). Little, Brown, Boston.

20 Vano, J.A., Udall, B., Cayan, D.R. *et al.* (2014) Understanding uncertainties in future Colorado River streamflow. *Bulletin of the American Meteorological Society*, **95**, 59–78. doi:10.1175/BAMS-D-12-00228.1.

21 Ritter, W. (2007) *Colorado Climate Action Plan: A Strategy to Address Global Warming.* URL http://www.colorado.gov/governor/images/nee/CO_Climate_Action_Plan.pdf [accessed on 9 September 2015].

22 Ray, A., Barsugli, J. & Averyt, K. (2008) *Climate Change in Colorado: A Synthesis to Support Water Resources Management and Adaptation.* URL http://wwa.colorado.edu/publications/reports/WWA_ClimateChangeColoradoReport_2008.pdf [accessed on 9 September 2015].

23 IPCC (2007) *Climate Change 2007: The Physical Science Basis. Contribution of Working Group I to the Fourth Assessment Report of the Intergovernmental Panel on Climate Change.* Solomon, S., Qin, D., Manning, M., Chen, Z., Marquis, M., Averyt, K.B., Tignor, M. & Miller, H.L. (eds) Cambridge University Press, Cambridge, United Kingdom and New York, NY, USA.

24 Lukas, J., Barsugli, J., Doesken, N. *et al.* (2014) *Climate Change in Colorado: A Synthesis to Support Water Resources Management and Adaptation, Updated Edition* edn. CIRES Western Water Assessment, University of Colorado Boulder, Boulder, CO.

25 Lowrey, J., Ray, A. & Webb, R. (2009) Factors influencing the use of climate information by Colorado Municipal water managers. *Climate Research*, **40**, 103–119.

26 Kaatz, L. & Waage, M. (2011) Denver Water's approach to planning for climate change. *Water Resources Impact*, **13** (**1**), 5–7.

27 Averyt, K. (2010) Are we successfully adapting science to climate change? *Bulletin of the American Meteorological Society*, **91**, 723–726.

28 Romsdahl, R.J. & Pyke, C.R. (2009) What does decision support mean to the climate change research community? *Climatic Change*, **95**, 1–10.

29 Meyer, R.M. (2010) *Public Values, Science Values, and Decision Making in Climate Science Policy.* PhD Dissertation, Arizona State University, Tempe, AZ.

CHAPTER 12

Evolving the practice of Regional Integrated Sciences and Assessments

Adam Parris[1], Sarah L. Close[2], Ryan Meyer[3], Kirstin Dow[4] and Gregg Garfin[5]

[1] Science and Resilience Institute at Jamaica Bay, Brooklyn College, 2900 Bedford Ave, Brooklyn, NY 11210, USA

[2] University Corporation for Atmospheric Research, in service to: Climate and Societal Interactions Division, NOAA Climate Program Office, 1315 East-West Highway, SSMC3, Silver Spring, MD 20910, USA

[3] California Ocean Science Trust, 1330 Broadway, Suite 1530, Oakland, CA 94612, USA

[4] Carolinas Integrated Sciences and Assessments RISA, Department of Geography, University of South Carolina, 709 Bull Street, Columbia, SC 29208, USA

[5] Climate Assessment for the Southwest (CLIMAS), School of Natural Resources and the Environment, Institute of the Environment, The University of Arizona, 1064 E. Lowell St., Tucson, AZ 85721, USA

12.1 Introduction

In the mid-1990s, scientific advances in climate science and a demand for context-specific climate information led the National Oceanic and Atmospheric Administration (NOAA) to embark on an experiment in regional climate assessments. The first team of the Regional Integrated Sciences and Assessments (RISA) program was formed in 1995, and the program grew over the next 20 years into the current network of 11 teams in regions around the United States. In this time, RISA teams grappled with the complexities of an enduring truth: science is but one small part of the decision-making process. The institutional, legal, political, and cultural contexts within which people use scientific learning are far broader than discrete data and information needs. The scientific approaches of RISA teams evolved in response to these contexts.

Diverse approaches used by individual RISA teams were critical not just because of the different regional settings, but also to advance the concept of integrated climate services. The stories included in *Climate in Context* are a powerful demonstration of the work conducted by the many RISA researchers over the years, and they are a small subset of this body of

Climate in Context: Science and Society Partnering for Adaptation, First Edition.
Edited by Adam S. Parris, Gregg M. Garfin, Kirstin Dow, Ryan Meyer, and Sarah L. Close.
© 2016 John Wiley & Sons, Ltd. Published 2016 by John Wiley & Sons, Ltd.

work. Owing to RISA success, other scientific programs and climate service providers are attempting to replicate these approaches and see diversity both as a virtue and as a drawback. In one sense, experimentation with diverse approaches allows for innovation. In another sense, it is more challenging to identify and replicate specific elements of RISA work that can lead to success.

Previous assessments of RISA and similar programs identify specific elements of successful decision-support efforts, such as forming connections across disciplines and sectors and linking decision-makers with existing information [1–3]. However, the thematic areas in *Climate in Context* frame integrative scientific objectives, within which individual approaches and elements can be tested and modified to suit complex contexts. Thus, *Climate in Context* advances the expansion of integrated climate services by positioning rigorous case studies of RISA work into themes, which can be applied to other scientific programs: Understanding Context and Risk, Managing Knowledge-to-Action Networks, Innovating Services, and Advancing Science Policy (Figure 12.1). In this final chapter, we illustrate how context constrains transferability and, as a result, why institutional capacity is critically important to successful climate service and adaptation efforts.

12.2   Embracing the complexity of context

RISA teams started as incubators for experimental climate services, many of which have led to adaptive decisions. Seasonal fire forecasts, drought visualization tools, and outlook forums are different forms of climate services developed by RISAs over the years (see Chapters 7, 8, and 9, respectively). These experiments are based on the specific approach of considering user needs early in the process of conducting research and developing decision-support tools. Through a long-term investment in social science, RISAs have built an understanding of complex and evolving decision-contexts, which substantially expands upon the foundation of identifying users' data or information needs. In this time of increasing demand for climate services, it is important to acknowledge the interplay of factors affecting climate-related decisions (e.g., legal, institutional, and cultural issues). These factors can make for a challenging context in which to incorporate operational climate service products, such as data or forecasts. Thus, to deliver these products in an effective way, and to make them usable, it is critical to understand the constraints under which decision-makers operate, the barriers to using new products, and the opportunities available to incorporate these products.

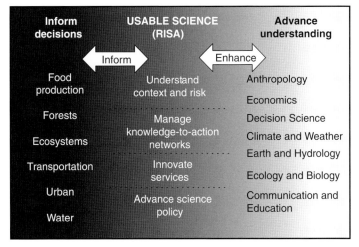

Figure 12.1 Conceptual framework for the RISA program. RISA teams integrate disciplinary expertise across the biophysical and social sciences within thematic areas (middle column) important to successfully informing climate-related decisions (middle to left column). These themes are connected nonlinearly; some of these steps may occur simultaneously, they may proceed linearly, or they may loop back on one another. Through their engagement with communities, social and behavioral scientists help understand the values, perceptions, and institutions that shape decision contexts (Chapter 1). These interactions are often catalyzed by information regarding socio-ecological risks posed by weather and climate. Services emerge as the tools and processes that help inform decisions in a timely and relevant manner. However, because the timing and relevance of decisions are acutely sensitive to context, the entire process is ripe for formal evaluation and assessment, which can in turn help to make approaches transferrable (middle to right column). Knowledge gained through decision-making experience becomes a central part of RISA work making the contrast between the scientific and decision-making communities less stark (all columns).

For example, consider that, for clean water to pour from your kitchen tap, a series of "upstream" decisions have to be made. Water utilities consider how to design infrastructure, such as billion-dollar dams or extensive delivery pipelines, and estimate the demand for water for your home, millions of other homes, and a myriad other uses (e.g., agriculture and fire suppression). These decisions are part of a fabric of context influenced by public perceptions, science and technology, political and legislative processes, and natural processes.

Timely and accurate seasonal forecasts can inform water managers about the availability of water in different places at different decision times within an annual calendar set by water policy (see Chapter 2). Seasonal forecasts require an informed scientific opinion based on an understanding, for example, of the variability of precipitation, changes in weather patterns that deliver precipitation (both rain and snow) and the reliability and persistence of precipitation in the future. Even with timely and accurate forecasts, water managers then have to make decisions about how to coordinate water releases among multiple reservoirs for multiple uses. Thus, the reliable delivery of drinking water that flows out of the tap in your home is based on what we know about climate, through scientific understanding as well as through management experience.

Climate services developed by RISA teams, in consultation or collaboration with service users, often stem from the ability, over time, to bring research and decision-making contexts into harmony. The capability to understand context and risk cannot be automated and may never be operationalized like forecast products or information services. This insight is a key reason why the RISA community continually stresses the importance of sustained and iterative engagement—it is the core part of a successful climate service. While there may not be shortcuts, there are common contextual factors to be considered, contended with, or in some cases harnessed in the design of climate services and usable science. The contributions to *Climate in Context* highlight some of these factors:

Policy implementation—Implementation of existing legislation sometimes reveals climate sensitivity and presents opportunities to involve science in decision processes. For example, Carbone *et al.* (Chapter 8) illustrate how the FERC re-licensing of dams in the Carolinas gave rise to the Dynamic Drought Index Tool, and Ray and Webb (Chapter 2) note the role that the Coordinated Reservoir Operations agreement played in creating an audience receptive to seasonal forecasts.

Climate-related hazards—Hazards, such as weather extremes and floods, often serve as focusing events. Hurricane Sandy (Chapter 3), the drought in the Southern Plains (Chapter 9), and wildfires in the Southwest (Chapter 7) are all examples where climate-related hazards created an opportunity for engagement and catalyzed a need for science to inform decisions.

Socio-economic and socio-ecological resource management—The diverse and evolving approaches to managing socio-economic and socio-ecological shifts in different communities can generate new science needs. Garfin *et al.* (Chapter 7) note how fire management in the U.S. has evolved in response to development patterns and changing demographics near areas prone to wildfire. Fire managers eventually saw a need for ongoing climate information and a dialogue around seasonal conditions, to inform resource allocation decisions. Schmitt Olabisi *et al.* (Chapter 5) demonstrate how exposure to heat waves varies depending on how emergency managers and public health officials can mobilize resources to different parts of cities, some of which are disproportionately exposed to the adverse effects of high heat.

Science and technology—Advances in scientific understanding and/or technology can create opportunities to improve climate services. Advances in seasonal predictions of climate variability in the mid-1990s led WWA to develop a series of iterative engagements and research with federal, state, and local government and other stakeholders, to advance the use and usability of seasonal forecasts, as described in Chapters 11 and 2. Trainor *et al.* (Chapter 6) document how the use of webinar technology has helped ACCAP increase awareness of climate-related risks in Alaska. Horton *et al.* (Chapter 3) explain the development of climate hazard profiles in New York City and the resultant ability of the Mayor's office to advance its long-term plans and policies.

These contextual factors evolve individually and in relation to each other. Decision contexts are also evolving as decision-makers and practitioners are increasingly considering climate in their decisions and operations. RISA teams often cite the need to allow for flexibility in approach because it is impossible to fully anticipate how these and other factors will interact to create opportunities for climate services. Deliberately responding to these factors and decision contexts helps to structure learning across various project- and program-level engagements and helps to set expectations when resources have to be shifted in response to an event (e.g., a large hurricane or intense drought). Moreover, it helps to manage transaction costs associated with engagement. RISA teams do not constantly proliferate and intensify engagement in response to each of these factors; rather, they are attentive to the need to be involved in ongoing dialogues and to provide timely information to different decisions.

One way RISAs minimize transactional costs is through staff and partners embedded in other communities, who can extend the reach and capacity of the RISA team. For example, a motivation for RISAs to fully integrate Extension staff into their teams is that Extension personnel are already well networked within agricultural and coastal communities. Stevenson *et al.* (Chapter 4) explain how RISAs can engage in and learn from these communities, while minimizing disruption to existing relationships and social networks.

Deliberately responding to contextual factors in the design and implementation of research is itself a reflection on broader social learning, specifically geared toward increasing climate resilience (see Chapter 10). Whether through purposeful design or serendipitous opportunity, the progress made toward improving the design of climate services suggests that RISA teams and related initiatives can strive to move beyond the goal of increasing the use of climate information toward fostering greater capacity to manage climate risks.

12.3   Emergence of evaluation

Pursuing societally driven goals and outcomes raises expectations for demonstrating the cause and effect of RISA efforts. As Ferguson *et al.* (Chapter 10) discuss, RISA's success is illustrated largely through case studies and anecdotal evidence. Through more formal evaluation efforts, RISA teams are beginning to test causal theories about how their response to contextual factors discussed above lead to societal outcomes such as climate resilience. Like Understanding Context and Risk, RISA evaluation efforts emerged from years of thoughtful experimentation and learning with cooperation between NOAA staff overseeing the RISA program and researchers on individual RISA teams.

Formal evaluation studies of RISA research emerge in the literature around 2005–2007 [4], well after the onset of the program. NOAA initiated the five-year competitive funding model for RISA around 2006. Proposal review has since provided a level of assessment, but not as rigorous as those methods described in Chapter 10. From 2006 to 2010, NOAA gradually encouraged evaluation in funding opportunities associated with the National Integrated Drought Information System's "Coping with Drought" initiative and competitions for RISA cooperative agreements (e.g., response by the Pacific RISA described in Chapter 10). During meetings of all the RISA teams in 2011–2012, NOAA program managers and RISA scientists extensively discussed the need to further develop evaluation capacity within RISA teams, and it has been a recurrent theme in each successive program-wide meeting. Today, approximately three-quarters of the RISA teams have a core member or, more commonly, an external evaluator specializing in evaluation.

Why detail this history? The development of formal evaluation illustrates collaborative governance or partnership between NOAA, RISA teams, and decision-making partners. The program has avoided over-emphasis of common evaluation tools, such as surveys and metrics, in favor of encouraging regionally-based evaluation researchers who interact directly with the teams. Evaluation approaches used by RISA teams, whether surveys, interviews, or other methods, yield rich results because they aim at incorporating regional context through direct engagement with the teams, thereby accounting for cultural, institutional, and political context. NOAA and RISA teams helped introduce evaluation researchers and their work to the climate science community by convening panels on evaluation at conferences geared to the physical science community as well as by NOAA's introduction of science policy into the RISA program objectives.

Having evaluation researchers participate in the partnership between NOAA, RISA teams and their decision-making partners helps in a number of ways. First, NOAA better understands where external contextual factors affect the impact of RISA work, for better or for worse. Second, NOAA incorporated insights from evaluation research into technical evaluation criteria for RISA funding opportunities. For example, applicants are encouraged to demonstrate a plan for engagement in proposals, as opposed to simply asking applicants who was engaged. Third, evaluation research supported by RISA informed the evaluation framework for the third U.S. National Climate Assessment (NCA) and a workshop on the evaluation of future NCA efforts.

Because contextual factors are critically important to RISA work, evaluation cannot be treated solely as a matter of bibliometrics or of counting the number of people engaged. The RISA community is working toward a more robust and broadly applicable set of outcome measures. However, the interplay between the complexity of context, the importance of sustained engagement, and the expertise required to discern what the scientific and decision-making communities can learn from the RISA program underscore a need for greater institutional capacity. RISAs are no longer pilot projects. They have evolved from their beginnings as university-based teams into a diverse set of adaptive networks in which partners from the public, non-governmental, and private sectors build capacity on both sides of the science and decision-making interface.

12.4 Conclusions

Climate in Context demonstrates clear success in addressing, on a local to regional scale, the central challenge of the U.S. Global Change Research Act of 1990—developing and providing "usable information on which to base policy decisions relating to global change" [5]. However, it also demonstrates the complexity of achieving what can be considered a straightforward, no-regrets strategy for any region—better preparing the public and our environment for weather and climate risks. Formal evaluation is central to learning how to advance the practice of RISA efforts, but it is not required to increase our confidence in usable science as a publically valuable investment. As our understanding of context grows, we have come to recognize that the design and evaluation of usable science require more attention to the nuances of context, and the richness of the decision-maker's environment.

The contributions herein help provide various strategies and lessons that can be built upon to meet the scientific challenges posed by this increasing complexity. To that end, we recommend: attention to context, including to the synergies among science, management, technology, and policy; purposeful engagement; and research that advances scientific understanding of climate and the factors involved in climate-related decisions. Throughout this process, these research and engagement strategies should be accompanied by meaningful and rigorous evaluation of the efficacy of processes to co-produce knowledge that informs decision-making.

The potential of the RISA program—the "voyage of discovery" [3,6]—is far from fully realized. Publicly-funded science should provide public value, such as informing reliable delivery of clean water to your home. Yet, scientific programs like RISA are few, and those are still viewed as pilot efforts among federally-sponsored programs. We have yet to transfer RISA practice back into the science system at a scale that fits the magnitude of the risks posed by weather and climate. Notable progress exists in the third U.S. National Climate Assessment (NCA), but even the ongoing NCA process struggles to cope with the competing needs of creating operational and easily replicable processes for learning, and the reality of place-based specificity, which augurs against a one-size-fits-all approach. Reflecting on this collection of RISA work, we envision the next frontier for RISAs and organizations with similar missions lies with designing and adaptively managing science to fit dynamic societal contexts. For this endeavor to successfully meet the full magnitude of the challenges posed by weather and climate, new publicly-funded science policies must be deliberately designed to balance a historic investment in fundamental understanding with new investment in knowledge tailored to societal context.

Disclaimer

The scientific results and conclusions, as well as any views or opinions expressed herein, are those of the authors and do not necessarily reflect the views of NOAA or the Department of Commerce.

References

1 National Research Council (2009) *Informing Decisions in a Changing Climate*. National Academies Press, Washington, DC.

2 National Research Council (2010) *Informing an Effective Response to Climate Change. America's Climate Choices*. National Academies Press, Washington, DC.

3 Pulwarty, R.S., Simpson, C.F. & Nierenberg, C.R. (2009) The Regional Integrated Sciences and Assessments (RISA) program: crafting effective assessments for the long haul. In: Knight, C.G. & Jager, J. (eds), *Integrated Regional Assessment of Global Climate Change*. Cambridge University Press, Cambridge, UK, pp. 367–393.

4 McNie, E.C. (2007) Reconciling the supply of scientific information with user demands: an analysis of the problem and review of the literature. *Environmental Science & Policy*, **10**, 17–38. doi:10.1016/j.envsci.2006.10.004

5 101st United States Congress 1990. *Global Change Research Act of 1990*. Public Law 101–606 [S. 169].

6 Miles, E.L., Snover, A.K., Whitely Binder, L.C., Sarchik, E.S., Mote, P.W. & Mantua, N. (2006) An approach to designing a national climate service. *Proceedings of the National Academy of Sciences*, **103** (**52**), 19616–19623.

Acronyms

Acronym	Definition
ACCAP	Alaska Center for Climate Assessment and Policy
ACF	Apalachicola–Chattahoochee–Flint
ACIS	Applied Climate Information System
ALM	Action-Logic Model
AOOS	Alaska Ocean Observing System
APRFC	Alaska Pacific River Forecast Center
ARCUS	Arctic Research Consortium of the US
BLM	Bureau of Land Management
CBRFC	NOAA Colorado Basin River Forecast Center
CCATF	Climate Change Adaptation Task Force
CCRUN	Consortium for Climate Risk in the Urban Northeast
CEFA	Climate, Ecosystems, and Fire Applications
CIG	Climate Impacts Group
CIRC	Climate Impacts Research Consortium
CISA	Carolinas Integrated Sciences and Assessments
CLD	Causal Loop Diagram
CLIMAS	Climate Assessment for the Southwest
CNAP	California-Nevada Applications Program
CoP	Community of Practice
CPC	Climate Prediction Center
CPO	Climate Program Office
CRI	Climate Risk Information
CROS	Coordinated Reservoir Operations
CRWCD	Colorado River Water Conservation District
CSIRO	Commonwealth Scientific and Industrial Research Organisation (Australian Government)
CSPM	Climate Services Program Manager
CWCB	Colorado Water Conservation Board

Climate in Context: Science and Society Partnering for Adaptation, First Edition.
Edited by Adam S. Parris, Gregg M. Garfin, Kirstin Dow, Ryan Meyer, and Sarah L. Close.
© 2016 John Wiley & Sons, Ltd. Published 2016 by John Wiley & Sons, Ltd.

Acronym	Definition
CWRM	State of Hawai'i Commission on Water Resource Management
DDIT	Dynamic Drought Index Tool
DOI	US Department of Interior
EIS	Environmental Impact Statement
ENSO	El Niño-Southern Oscillation
ESA	Endangered Species Act
FAR	IPCC Fourth Assessment Report
FEMA	Federal Emergency Management Agency
FERC	Federal Energy Regulatory Commission
FRCCG	Front Range Climate Change Group
GACC	Geographic Area Coordination Center
GCM	Global Climate Model
GHG	Greenhouse Gas
GINA	Geographic Information Network for Alaska
GLISA	Great Lakes Integrated Sciences and Assessments
IARC	International Arctic Research Center
ICAP	University of Hawai'i Center for Island Climate Adaptation and Policy
IDF	Intensity–Duration–Frequency
IPCC	Intergovernmental Panel on Climate Change
IWCD	Intermountain West Climate Dashboard
IWCS	Intermountain West Climate Summary
KAN	Knowledge-to-Action Network
KBRR	Kachemak Bay National Estuarine Research Reserve
KML	Keyhole Markup Language
LCC	Landscape Conservation Cooperative
LiDAR	Light Detection and Ranging
LNDT	La Niña Drought Tracker
M&Is	municipal and industrial water utilities
MDCH	Michigan Department of Community Health
MDEQ	Michigan Department of Environmental Quality
MEC	Metropolitan East Coast
MMHM	Mid-Michigan Heat Model
NACSP	North American Climate Services Partnership
NAME	North American Monsoon Experiment
NASA	National Aeronautics and Space Administration
NCA3	Third National Climate Assessment

Acronym	Definition
NCDC	National Climatic Data Center
NDMC	National Drought Mitigation Center
NEPA	National Environmental Policy Act
NFP	National Fire Plan
NGO	Non-Governmental Organization
NICC	National Interagency Coordination Center
NIDIS	National Integrated Drought Information System
NIFA	National Institute of Food and Agriculture
NIFC	National Interagency Fire Center
NJ	New Jersey
NOAA	National Oceanic and Atmospheric Administration
NPCC	New York City Panel on Climate Change (NPCC1 = first NPCC, NPCC2 = second NPCC)
NPS	US National Park Service
NRC	National Research Council
NSAW	National Seasonal Assessment Workshop
NWS	National Weather Service
NY	New York
NYCDEP	New York City Department of Environmental Protection
NYS	New York State
OLTPS	Office of Long Term Planning and Sustainability
PI	Principal Investigator
PIRCA	Pacific Islands Regional Climate Assessment
PNG	Portable Network Graphics
PSD	Physical Sciences Division of NOAA's Earth System Research Laboratory
PS	Predictive Services
RCC	Regional Climate Center
RCES	Regional Climate Extension Specialist
RCP	Representative Concentration Pathways
RCSD	Regional Climate Services Director
RECS	Regional Extension Climate Specialist
RISA	Regional Integrated Sciences and Assessments
SCDNR	South Carolina Department of Natural Resources
SCIPP	Southern Climate Impacts Planning Program
SECC	Southeast Climate Consortium
SERI	Social and Environmental Research Institute

Acronym	Definition
SIRR	Special Initiative on Rebuilding and Resiliency
SRES	IPCC Special Report Emissions Scenarios
SVG	Scalable Vector Graphics
SWCO	Southwest Climate Outlook
SY	Sustainable Yields
TWG	Climate Change Technical Work Group
UACE	University of Arizona Cooperative Extension
UAF	University of Alaska Fairbanks
UKCIP	UK Climate Impacts Program
US	United States of America
USDA	US Department of Agriculture
USFS	US Forest Service
USFWS	US Fish and Wildlife Service
USGCRP	United States Global Change Research Program
VCAPS	Vulnerability and Consequences Adaptation Planning Scenarios
WUI	wildland–urban interface
WWA	Western Water Assessment

Index

Climate in Context: Science and Society Partnering for Adaptation, First Edition.
Edited by Adam S. Parris, Gregg M. Garfin, Kirstin Dow, Ryan Meyer, and Sarah L. Close.
© 2016 John Wiley & Sons, Ltd. Published 2016 by John Wiley & Sons, Ltd.